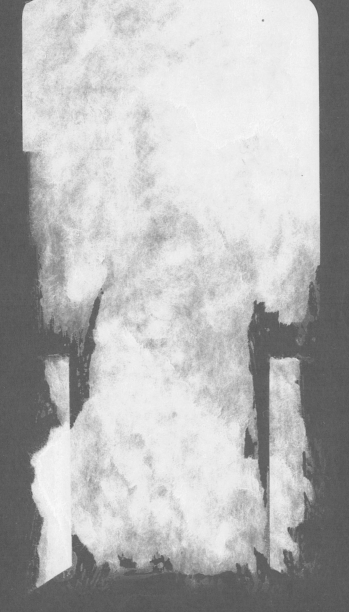

# Wind-Catchers

# Wind-Catchers

## American Windmills

## of Yesterday and Tomorrow

## Volta Torrey

The Stephen Greene Press

Brattleboro, Vermont

# Picture Credits

*L E*

The author is grateful to the institutions and individuals who supplied illustration material for WIND-CATCHERS, as follows: for pictures on **pages 4, 189**—Nebraska State Historical Society; **page 5**—Colonial Williamsburg; **pages 6, 76, 86, 91, 105**—The Smithsonian Institution; **page 7**—Grant Heilman, Lititz, Pennsylvania; **page 8, 178, 203**—National Aeronautics and Space Administration; **page 18**—The M.I.T. Press (Reprinted from Hans Wulff: *Traditional Crafts of Persia* by permission of The M.I.T. Press, Cambridge, Massachusetts); **page 27**—Lent to Science Museum, London, by Kings College, University of London; **page 31**—North Dakota State University, Fargo; **page 41**—British Crown Copyright, Science Museum, London; **page 53**—Copyright Doeser Fotos, Laren, N.H.—Holland; **page 55**—Joel's Studio, Holland, Michigan; **page 67**—Newport Restoration Foundation, photo by Jan Armor; **page 82**—Heritage Plantation of Sandwich, photo by William A. Newman; **page 87**—Library of Congress; **page 93**—Windmill Gardens, Fremont, Michigan; **page 109**—Union Pacific Railroad; **pages 133, 135, 137**—Don Guy, Lincoln, Massachusetts; **page 149**—Mrs. Eleanor Rossi Crabtree, San Francisco; **page 154**—School of Engineering and Applied Sciences, Princeton University; **page 158**—*Popular Science*; **page 160**—Wind Power Systems; **page 181**—Sandia Laboratories; **page 182**—Alberto Kling; **page 196**—Pennawalt Corporation; **page 197**—Dyna Technology; **page 205**—Photo by Dev O'Neill.

This book has been produced in the United States of America:
designed by R. L. Dothard Associates, composed by Publishers Composition Service, and printed and bound by Halliday Lithograph Corporation.

It is published by The Stephen Greene Press, Brattleboro, Vermont 05301

LIBRARY OF CONGRESS CATALOGING IN PUBLICATION DATA

Torrey, Volta, 1905–

   Wind-catchers: American Windmills of Yesterday and Tomorrow.

   Includes bibliographical references.
   1. Windmills—United States—History.   2. Windmills—History.   I. Title.
TJ823.T67        621.4′5        76-13816
ISBN 0-8289-0292-5

PUBLISHED NOVEMBER 1976
*Second printing March 1977*

THIS BOOK IS FOR
AN INDEFATIGABLE TEACHER
WHO IS MY WIFE

# Contents

# Acknowledgments

T H I S  B O O K was the publisher's idea, not mine. "At the time you started it," a close friend said recently, "I thought you were a bit balmy—that nobody would buy a book on windmills. Now I suppose anything on energy goes big."

That friend, Edward Morrow, was one of many persons who helped collect widely strewn bits of information about a charming old engine that many able historians seem never to have noticed. Robert M. Vogel, the Smithsonian Institution's Curator for Civil and Mechanical Engineering, permitted me to search his personal files as though they were my own. At the Library of Congress, Jane Collins kindly kept a desk covered with source material I needed. Reference librarians at the Folger Shakespearean Library and the Martin Luther King Library in Washington guided me to unsuspected sources. Historical society staffs in several states—especially Nebraska, New York, North Carolina and Indiana—answered many questions.

At NASA Headquarters, Frank Rowsome, Jr., Howard Allaway, Sandra Scaffidi, Grace Reeder, Victor Coles, H. Kurt Strass, Lee O. Saegesser, Nancy Ebert, et al, volunteered many suggestions. At M.I.T., Professor Bud Wilbur, John Mattill, and other friends and acquaintances explained technical problems to me.

Sympathetic helpers elsewhere included Edward Hutchings, Jr., at Caltech; Bill Pinkerton at Harvard; Hubert Luckett and Bill Morris of *Popular Science Monthly*'s staff, and former Associated Press reporters William McGaffin, Joe Wing, and Don Guy. I am likewise indebted to many personal and professional friends including Eric Sloane, Harland Manchester, Richard Dempewolff, Jack Stearns, Robert Duffy, Agnes Robinson, Pauline Stanley, Marcellus Henderson, and Lenore Daly.

Newspapermen and other writers whom I have never met sent me clippings and answered questions about old and new windmills; Ted Brook of the *Wichita Eagle*, Dolph C. Simons, Jr., of the Lawrence (Kansas) *Journal-World*, Whitney Gould of the Madison (Wisconsin) *Capital Times*, Dick Hellner of the McCook (Nebraska) *Daily Gazette*, and Tom Allan and James Denney of the *Omaha World-Herald* were especially kind.

Many other persons to whom I am grateful are named in the text and references. Mr. and Mrs. Stephen Greene provided pictures to illustrate the text. I am wholly responsible for the opinions expressed, some conclusions, and the errors that better informed readers will probably find in this review of facts and fiction about the flirtatious machines called windmills.

*Washington, D.C.*
*May, 1976*

# A Condensed Wind Engine Chronology

900s: Horizontal wind wheels irrigate Persians' gardens.

1200s: Vertical sails on posts and towers grind grain for European land-lords and peasants.

1500s: Windmills begin to reclaim lowlands in Holland; Don Quixote attacks them in Spain.

1600s: Colonists build European types of windmills on America's eastern seaboard.

1700s: Steam engines begin to replace wind-driven wheels.

1860s: Americans produce multi-bladed fans to run pumps.

1870s: Chicago becomes the hub of the windmill industry.

1880s: Windmills help to make a Great American Desert bloom.

1900s: Dutch windmills beautify a San Francisco park; the Danes generate electricity from the wind.

1920s: Tiny windmills light up and enliven rural homes in America with radios.

1930s: Central power plants and transmission lines replace windmills.

1940s: Vermont has most powerful wind-driven turbine ever built.

1950s: Atomic energy diminishes interest in using the wind.

1960s: Persistent pollution alarms millions of Americans.

1970s: Fuel scarcity revives interest in the wind's energy.

1976 : United States begins first multi-million dollar research and develop-ment program to convert the wind's power to kilowatts.

I

# *Why Windmills?*

*Falstaff*: What wind blew you hither, Pistol?
*Pistol*: Not the ill wind that blows no man to good.

HENRY IV, PART 2, ACT FIVE SCENE 3

FROM THE MOON the earth would look different every time you glanced up at it because the wind here never rests. Somewhere in our sky it rearranges the clouds every minute, like an artist forever trying to make a picture more beautiful. People have used wee bits of the wind's inexhaustible energy for thousands of years, some still do, and many more may before long. When I was a boy in Pottawatamie County, Iowa, a grumbling old windmill sometimes lulled me to sleep at night, and I often played in its wooden tower as if it were a jungle gym. That patient machine above the barnyard pump has vanished, however, along with tens of thousands of other farm windmills. Most school boys now learn as much about atomic energy as about aeolian energy.

*Aeology* is the branch of physics that deals with the atmosphere, the life-blood of our planet. Until recently practical aeologists tended to be more interested in ways to divert or block the wind than they were in ways to make it serve useful purposes. Is this because men have learned so much about the wind, or because they have forgotten what inventive people learned long ago?

1 ঌ

The wind can scatter the leaves we rake together on our lawns in a jiffy, but it takes millions of years for nature to turn a heap of organic matter into a vein of coal or a pool of oil with stored-up energy that we can use. The Persians who built windmills to irrigate their gardens ten centuries ago did not suspect their treasures underground. They did not suspect that the sun derives its energy from nuclear reactions, or that the radiation from the sun drives the wind from place to place on our whirling planet orbiting the sun. But the Persians knew energy when they saw it, and used their heads to save their strength.

Until energy shortages began to frighten Americans, many had belittled windmills. "Too primitive . . . unreliable and inefficient," a steel company's advertisement said of them, and many of my friends laughed when they heard I was writing a book about windmills. Few persons realized how precious fossil fuels were becoming until we were asked to turn off the tiny lights on our Christmas trees in 1973.

By then, the ever rising demand for electrical energy had already prompted our government and private corporations to invest huge sums in nuclear power plants. Although the plants are still controversial, the research that led to their development has been valuable. Technology developed for fission and fusion reactors may help engineers exploit other sources of energy—including the wind.

At Princeton University physicists have been building colossal contrivances to fuse nuclear particles. They hope to produce electricity by the process used in a hydrogen bomb. If they succeed, mankind may never need any other source of power. But they haven't gotten far yet. On the same campus, meanwhile, a new kind of wind wheel began to generate electricity in the 1970s. When science reporters asked a director of the fusion research project about it, he assured them that the windmill was just "a toy some of the boys had built," perhaps forgetting that his own laboratory was full of instruments that were once toys, too.

From discoveries of all kinds engineers are learning more than anyone knew before about properties of structural materials, properties that have been serious limitations on the size of wind-driven generators. From space exploration especially, men have learned more about storing electrical energy, which is now a costly process that has helped make wide use of the wind's power in our homes expensive. When these and related problems are minimized, the wind may again be as helpful as it was in Holland in the sixteenth century and on our own country's Great Plains in the nineteenth.

### ENTERTAINING ENGINES

Like sailing ships, windmills have endeared themselves to more generations of mankind than have engines that smoke. Instead of darkening the sky, windmills have made forlorn parts of the world more beautiful. Tens of thousands of them are still running, and more may rise sooner than seemed likely only a decade ago.

Most of us call every kind of wind-driven engine a wind-*mill*, because for centuries the most important use for the engines was to grind grain into flour. Manufacturers emphasize the differences between wind engines built to turn millstones, pump water, and generate electricity. Historians, architects, and engineers divide windmills into many more categories. The wind, however, ignores nomenclature and makes all wind-driven engines do what comes naturally. They obey the wind gladly, stop working when not driven, and have to be kept from running wild when the wind's pressure on their sails becomes excessive. Reviewing men's efforts to get more work out of them is fascinating. Nearly every imaginable trick seems to have been tried.

When converted into mechanical or electrical energy, the kinetic energy in the wind can be used like the chemical energy in fossil fuels. Engineers measure the density and velocity of the amount of wind intercepted per second or hour by a windmill's sails to compute the power obtainable from them. The volume intercepted has always been small, and only minute fractions of the wind's power have ever been put to use.

Wherever men could, they have exploited water power instead of windpower, because water is easier to store, channel, and manage. Water turbines can generate electrical energy more steadily and reliably than turbines dependent on the wind. The air is a flirtatious force, thinly distributed, and the wind is often strongest when and where we need it least. Even with the help of very fast electronic computers, forecasters cannot be certain what the wind will do at a

3 ε∾

particular point in time or space. After the National Weather Service failed to predict a severe storm in Washington, D.C., a meteorologist explained that, even if every relevant fact were fed into the Weather Bureau's fastest computer, the weather might change before the forecast could be issued.

In spite of the wind's unpredictability, however, people have persisted in trying to harness its energy, and have often been rewarded generously. In this book we will recall the origins of five big families of windmills—horizontal, post, smock, American, and aerodynamic machines—and note some interesting marriages among them.

*"Merry-Go-Round" pumping water near Lincoln, Nebraska in 1898.*

**Horizontal rotors:** The Persians had the first windmills that much is known about. Their machines were built by attaching sails to center posts to be whirled around horizontally by wind from any direction, like horses on a carousel. Less than a century ago many such homemade merry-go-rounds ran pumps in the United States, and some of the newest types of windmills being tested today have horizontal rotors.

~§ 4

*The reconstructed Robertson windmill on a post at Williamsburg, Virginia.*

**Post mills:** In northern Europe millwrights got more power from the wind than the Persians had. They did this by intercepting the wind's path with nearly vertical sails. Those sails rose and fell majestically, but had to be kept facing the direction from which the wind came, and had to be furled when the wind became ferocious. This was managed at first by pivoting whole millhouses on posts. When this was done the structures were often so beautiful that replicas of old post mills are being built to enable more people to enjoy seeing them. On many new machines electrical generators are now being pivoted similarly on high towers.

5 ૐ

*A smock mill near Yarmouth, Cape Cod, resembles those built in Europe.*

**Smock mills:** Few post mills have survived time's wear and tear, but many smock mills have. They were structures developed long ago on which only the caps had to be turned to redirect the sails when the wind veered. In Greece the sails were flexible fabrics like the canvas on some small boats, but in northern and western Europe the sheets were usually held taut on the ribs in wooden frames. Smock mills could be taller and plainer than post mills, and some have been well preserved on America's eastern seaboard.

6

*Typical American farm windmill. Many like it are still running.*

**American windmills:** Rotors with many more and much shorter blades than European windmills usually had to intercept the wind were invented in the United States in the 1860s. They were not as pretty as older European types, but they were easier to produce and more efficient. Factories in our midwestern states built and sold millions of these "pinwheels on stilts" until about half a century ago. The railroads used them to fill water tanks for locomotives, rich men bought them to have running water in their homes, and ranchers and homesteaders used them to irrigate our once arid Great Plains.

7 ठ≱

NASA *put aerodynamic sails on this turbine in Ohio.*

**Aerodynamic engines:** These have airfoil sails designed as scientifically as airplane propellers, and are used mainly to generate electricity. Their rotors, which usually contain only two or three blades, revolve faster than the older types of fans. The most powerful aerogenerator of this type ever built was a colossus that ran in Vermont in the 1940s until it lost one of its two blades. Nothing to equal it has been constructed anywhere since, though the United States may soon have a prototype of an even more powerful wind-driven engine. The discovery of atomic energy resulted in a hiatus in windmill research and development in this country during and after the Second World War. Stewart Udall, a former Secretary of the Interior, once complained that "As far as I know the government is not spending any funds on wind-energy research—a hell of a note if you ask me." But sensible citizens are reminded of windmills as the air becomes more foul and the energy crisis more acute. While head of the National Science Foundation, Dr. H. Guyford Stever pointed out that we live in "an energy thicket," and government agencies began to support studies of aeolian energy to determine its usefulness.

### WINDMILL WATCHING

Thanks to these developments, watching windmills may soon be as common a pastime in the United States as it is abroad. Tourists in the Netherlands and other European countries can buy guidebooks that list and describe historic windmills. Similar aids to recognizing them may soon be available here, but most of our country's noteworthy windmills are still only locally famous. This makes looking for them today like searching for rare fauna or flora in a jungle.

Windmill watching is a sport that everyone in your family can enjoy. Wherever you go you may discover a windmill. Museums and parks from Cape Cod to California have examples of various types. Portraits of them by noted artists hang in the great galleries. Historical societies exhibit models, photos, and remnants of windmills, and you find them running in amusement parks and campgrounds.

In this country, the best places to look for specimens of the European types of windmills are mostly near the Atlantic Coast, and the American type is most numerous on the central prairies—but there are examples of all types nearly everywhere. Big old horizontal windmills are the hardest to find, but there is one in northern New York and another in Nevada. Numerous new experimental wind generators that revolve horizontally have been set up outside great research laboratories.

At Williamsburg, Virginia, you can see corn meal ground between millstones in a post mill the way it was done for fastidious colonial cooks, and you can buy a bit of it. On Long Island and Cape Cod there are still smock mills similar to those that once pumped sea water into vats to evaporate and deposit salt. One of the finest tower mills ever built in the Netherlands was moved to Holland, Michigan, a few years ago. In San Francisco's Golden Gate Park, two even more powerful Dutch windmills await restoration of their sails. Danish windmills whirl gaily at Solvag, a few miles from Santa Barbara, California; and in Elkhorn, Iowa, money has been raised to have a 127-year-old windmill brought from Denmark.

Although a half million windmills were reported to be still upright a few years ago in the United States and Canada, many were in sad shape. Some windmill towers had been put to other uses. There was even a "Church in a Windmill" at a camp near Silver Lake in East Windham, New York. Windmills are sometimes converted into permanent homes in this country, as in Europe.

Many of the old American windmills are still at work on the Plains and farther west. In the rolling hills and dry valleys of central California, Mervel Tucker, a windmill mechanic at Shandon, has kept about two hundred machines in good running order, and has put up a few new ones for ranchers every year. On a hundred-mile drive through the scenic country south of Tucson, William Morris, an alert photographer, recently counted fifty American windmills. Dr. Vaughn Nelson of West Texas State University watched for them between Ashland, Kansas, and Perryton in the Texas panhandle, and saw 251, all but about a dozen of which were still usable.

Some of the newest windmills in our country were imported from Australia, Germany, Switzerland, France, and Argentina. Men handy with tools have also built their own; plans and instructions for doing this are readily available. Aerodynamic windmills are being made out of plastics as well as wood and metal.

From coast to coast, a windmill watcher soon learns to watch out for decoys. Merchants and showmen have littered the countryside with imitations of Dutch windmills. In many of these, like one at Disney World in Florida, the sails disregard the wind because they are spun by a hidden electric motor. On Cape Cod some new homes and guest houses have been intentionally built to resemble the octagonal windmill towers constructed two centuries ago.

The most deceptive of all decoys, no doubt, was the American type of windmill that the California Department of Public Works erected in 1972. There had been a surprising number of accidents on Interstate 5 south of Bakersfield. The accidents were ascribed to drivers' drowsiness, and state officials hoped that the sight of a windmill would wake people up. At first glance, the machine that was erected looked exactly like an old-fashioned farm windmill. But there were just as many accidents after the windmill was put up as before— possibly because it was too plainly a phony, driven by an electric motor, and lit at night by spotlights.

### FORTUNE HUNTING

Scanning the genealogy of today's wind-driven devices convinced Paul Harvey, a syndicated newspaper columnist, that "the windmill has not been discovered." He may be right. The windmills built thus far may be mere points of reference from which much better ways of "mining" the atmosphere for energy will soon be discovered.

To young men and women trained in systems engineering, many of the problems involved in making improved windmills seem closely allied to those encountered in designing aircraft and spacecraft. At the California Institute of Technology, where the faculty and students enjoy grasping such technological nettles, a course on windmills was added to the curriculum in the spring of 1974. Windpower, a professor explained, "may be a big business in the future," and two dozen students elected to take the new course. Both Dr. Ernest E. Sechler and Dr. Homer Joseph Stewart, who taught the class at Caltech, were distinguished aeronautical engineers and, Dr. Stewart was one of many specialists who helped Palmer Putnam design Vermont's wind-driven generator, the biggest that ever fed current into a modern utility system (see Chapter XII). Professor Hurd Willett of the Massachusetts Institute of Technology, who was also associated with Putnam's experiment, estimated at the time that the energy potentially available at good sites for wind-driven plants in the United States amounted to about a million megawatts. Many more watts may be drawn from the wind by more efficient methods that are currently being tested.

Dr. Stewart of Caltech believes that the wind's potential for energy production is several times that of the electrical plants serving this country today. The big question is, the professors agree, Can it compete economically with other power sources? What we pay for electrical current already is partly determined by the cost of building plants to serve large distribution systems. Aerogenerators have always cost too much to construct to interest many utility companies. They have found it more economical to build plants burning fossil fuels. But as those fuels have become costlier and more people have become concerned about the environment, efforts to develop alternative sources of power have been intensified.

It is partly as a result of such developments that Caltech students are being encouraged to think about windmills. With new aerodynamic blades to intercept the wind, composite materials that weigh less and are stronger than more familiar materials, better automatic controls, and more versatile electrical apparatus, it may be possible to generate much more current from the wind in the future than we can today. Wherever wind has a chance to develop great strength—on islands, open plains, smooth coastal shores, and in deserts and mountain passes—it may soon run many generators.

From space exploration we have learned how thin and fragile an atmospheric membrane protects our planet. We have polluted much of that atmosphere by burning coal and oil, and nuclear plants might add even more dangerous debris than smoke and soot. Some men now think there may be a pot of gold waiting for the developer of an efficient, inexpensive machine to generate storable, transportable energy from the wind—this despite the fact that the windmill inventor's only reward has too often been merely a card in a patent office file or a vote of thanks.

When I began working on this book I hoped to tell you how to build your own aerogenerator. Today so many new types are being tested that it is hard to say what kind anyone should build or buy. The windmill's whole history has been a comedy of errors; the machines have survived so many vicissitudes that they will surely live happily forevermore—after a few more troubles.

# PART ONE

# An Engine's Childhood

II

# Orphan of the Storms

'Tis all the music of the wind, and we
Let fancy float on this aeolian breath.
J. B. BRAINARD IN "MUSIC"

"ONCE UPON A TIME" is the best date that any-
one can give for the construction of the first windmill. No one knows
who first thought of using the wind for purposes other than scenting
food, foes, or friends, nor can historians say how, when, or where it
happened. The idea may have occurred to some *Homo erectus* idly
watching the wind rip the foliage off a tree or bush. Even before
men hoisted sails on boats, children no doubt romped in the wind,
and the first wheel made for it to spin may have been something like
the pinwheels we still make for children. Some people also have
speculated that primitive wind wheels may have been used for gam-
bling devices.

Buddhist monks in Tibet have watched prayer wheels turned by
the wind for twenty-five centuries. Everywhere people must often
have prayed for relief from tiresome chores. In ancient Egyptian,
Greek, and early Christian temples the words of prayers are said to
have been inscribed on cylinders hung from the ceilings. Seeing a
breeze turn one of those cylinders could have inspired someone to
make a wind engine. Whoever first made one found an answer to
many generations' prayers.

Beasts, slaves, and women ran Greek and Roman mills until men discovered that a flowing stream of water could do it. "Ye maids who toiled so faithful at the mill," Antipater of Thessalonia wrote in the first century B.C., "Now cease your work and from those toils be still." The Norse, too, built tiny water mills that ground exceedingly slow, and that were not improved for centuries. A crofter in Scotland explained, "If I get all the power I need from the burn as it flows past, where is the foolishness in leaving the rest unused?"

Historians suspect that the wind was first harnessed to run pumps in the Near East. Many authorities doubt wind-driven pumps were invented before the birth of Christ, though an emperor of Babylon, Hammurabi, was reported to have planned to use windmills in an ambitious irrigation scheme about 2050 B.C.

One of the world's most famous paintings, "The Procession to Calvary" by Peter Brueghel the Elder, shows a huge Dutch windmill on a high pinnacle at that memorable scene. The people in the procession seem indifferent to it, and it is extremely unlikely that they passed any such towering machine. Some viewers of that sixteenth century painting, now in the Vienna Museum, have, nevertheless, written that it symbolizes the birth of democracy and a machine age.

The Greeks included the four winds—Boreas, Eurus, Notus, and Zephyr—among their gods, and enjoyed the music produced by boxes of wind whistles. A drawing ascribed to Hero of Alexandria in the first century A.D. shows a wheel with sails around it pumping air into an organ. At first glance this is a puzzling picture because the organ has no keyboard. Hero may have intended to add one later, of course, or may have simply wanted to hear the wind blow whistles for the same reason that scientists now listen to whales and stars.

The wind still may be musical. A few years ago Ward McCain, a student of musical instruments, built a big harp in Vermont for the wind to play. He suspended its 80 stainless steel strings from a 25-foot oak beam and constructed a large wooden sounding box. The music from this novel instrument varied in pitch and volume as the wind rose, fell, and danced around it.

In Portugal long ago the millers tied ceramic pots to the arms of their windmills to make them wail like ghosts in modern horror movies. Those *jarras* or *buzios* were there to tell a miller, by the pitch of the sound they made, how fast or slowly his wheel was turning.

Women ran the mills while their husbands worked in the fields, and the sound could bring a man running to help his wife if the wind began to whirl the sails so fast that she could not stop them in time to save them.

By Charles Dickens' day barrel organs that automatically played certain tunes when air was pumped into them were popular. Near his home there was one in a windmill that Dickens liked to hear when the miller let the wind run the pump. In the United States, a Guild of Former Pipe Organ Pumpers flourished in the first half of this century. Its members were men who remembered having to pump air into church organs, and enjoyed telling each other what happened when they fell asleep or played pranks during long church services. The wind could have been even more mischievous than those boys. It is a blind workman who whistles while he works.

### PERSIAN WINDMILLS

The Arabs who felt the wind sweeping across the deserts where Mahomet preached seem to have recognized its power sooner than their contemporaries in pleasanter parts of the world. A sermon quoted by Mahomet's disciples suggests that he viewed the environment in somewhat the same way that some engineers do today:

"When God created the earth it shook and trembled, until he put mountains on it, to make it firm. Then the angels asked, 'Oh, God, is there anything of thy creation stronger than these mountains?' And God replied, 'Iron is stronger than the mountains; for it breaks them.' 'And is there anything of thy creation stronger than iron?' 'Yes, fire is stronger than iron, for it melts it.' 'Is there anything of thy creation stronger than fire?' 'Yes, water, for it quenches fire.' 'Oh, Lord, is there anything of thy creation stronger than water?' 'Yes, wind; for it overcomes water and puts it in motion.' 'Oh, our Sustainer! Is there anything of thy creation stronger than wind?' 'Yes, a good man giving alms; if he give it with his right hand and conceal it from his left, he overcomes all things.' "

Two tenth century Arab writers, Al-Mas'udi and Al-Istahri, mentioned windmills in Seistan, a high dry land near the present boundary between Iran and Afghanistan. "There the wind drives mills and raises water from the streams, whereby gardens are irrigated," Mas'udi

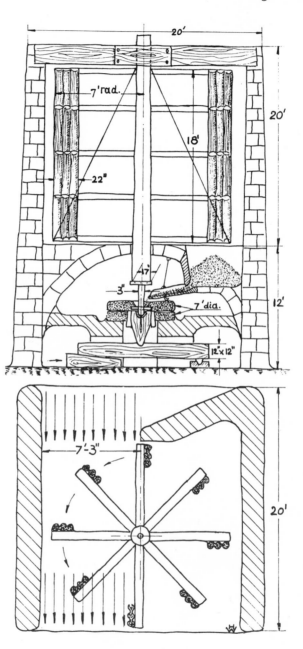

*Drawings show how Persians built windmills in Seistan.*

reported. "There is in the world, and God alone knows it, no place where more frequent use is made of the winds." Little is known about the origin of these machines.

Alexander the Great might have come across Persian windmills while fighting his way to the frontiers of India in the fourth century B.C., and prisoners of war could have spread knowledge of windmills from one country to another. A man forced to plod around in circles like a beast far from his home to pump water or turn a millstone would certainly have been tempted to show his captors how to let the wind do the work. The legends differ. By one account, the Caliph Omar asked a Persian slave if he could build a windmill and the man replied: "By God, I will build this mill of which the world will talk." By another account, the slave considered it so impossible that he killed the caliph.

In northwestern Europe windmills began to spring up after the Crusades. Christians may have returned with the Persians' know-how, although Seistan, where windmills were reported before the time of the Crusades, was not on the main routes to the Holy Land. In any case, some of the Crusaders may already have known how to build windmills, for an eyewitness to their great journeys wrote that "The Germans used their skill to build the first windmill that Syria had ever seen."

The Crusades awakened people in many lands. Art and architecture were affected. From then on the lords and ladies in northwestern Europe could "taste the spices of Arabia yet never feel the scorching Sun that brought them forth." Peasants felt less tightly bound by tradition's chains. "The exploratory character of western technology," Professor Lynn White, Jr. believes, accounts for the sudden popularity of windmills throughout Europe in the twelfth century.

The sails of most of the machines built in the north differed from those used in the south. This was no doubt because of the severity of the climate, the wind's characteristics, the materials available, and the workmen's skill. On the shores of the Mediterranean the wind's power could be tapped with light linen sails, and a mill did not have to be moved to another site as often as turned out to be desirable elsewhere. The frigid wild winds near the North Sea necessitated the construction of sturdier types.

From Greece to Portugal windmill sails were often triangular sheets of cloth strung from light masts supported by stone towers. Tourists in those countries still marvel at the longevity and numbers of those machines, and especially admire the dark red sails favored in some countries. Farther north, as post mills evolved, whole mills were pivoted on single wooden legs so that they could be turned to face a breeze from any direction. Later, in smock mills, only the caps or bonnets of the towers had to be turned when the wind's direction changed. Much more powerful windmills were constructed then and used for more purposes.

Nearly everywhere in Europe wind-driven wheels were first used mainly to turn millstones. Those stones were pairs of flat, thick, heavy slabs several feet wide. One was held stationary while a mate, called the "runner," was revolved above it. The grain was fed between them through a hole in the center of the runner, to be torn apart by furrows cut in the faces of both stones. The grist then was thrown out into a box around the lower, or "bedstone." To grind the grain properly, the runner had to be kept a proper distance above the bed-stone and the grain allowed to flow between the stones at a rate determined by the speed of the wind. Frequently, too, the runner had to be lifted off so that the furrows in the stones could be cleaned, and from time to time the grooves had to be carefully sharpened. Many men mastered those basic skills by running and maintaining water mills. But to run a windmill efficiently, the miller also had to become a good judge of the weather, alert even to small changes in the wind's strength, direction, and mood.

Don Quixote's error in attacking windmills in the sixteenth century was more plausible than most readers now realize. The "thirty or forty monstrous giants" that he encountered could have appeared in La Mancha while he was absorbed in his studies of chivalry. They were squat stone towers, with sharply pointed round roofs that resembled helmets, twirling mighty arms that nearly swept the grasses off the Spanish plain. From a distance they must have looked truly menacing to anyone seeing them for the first time.

In the Netherlands the first windmills looked friendlier. They were little wooden boxes on posts with arms that rose and fell like those of children twirling ropes. Tailpoles that extended to the ground behind each box might be mistaken from a distance for the tail of some

*Portuguese windmills had alarm bells between the arms.*

21  કે

enormous bird. The profile of a post mill on the horizon evoked even pleasanter thoughts in many imaginative Dutchmen's minds. They compared their country's mills to farmers' sturdy daughters, with square shoulders and broad skirts tightly bound to narrow waists, diligently toiling in their fathers' fields. Don Quixote would have doffed his hat and bowed to those civilized windmills.

## WIND RIGHTS AND TAXES

Despite the windmill's winning ways, it was not welcome everywhere. In feudal times the landlords owned whatever water flowed onto their property, and claimed the wind, too. They were obliged to let the peasants' grain be ground in their water mills, and collected a fee for that service. In medieval England, the church owned huge tracts of land on which there were many water mills. It used the profits from those mills to support its struggles to save men's souls. One churchman resisted the windmill's debut on that blessed island almost as fiercely as Don Quixote fought one in Spain, and as vainly.

Abbot Samson was so powerful that he dared say nay even to the king. When a windmill appeared on the abbot's land, he ordered it demolished, lest townsmen take their grain to the windmill rather than to the water mills from which the abbot collected a fee. The bishop responsible for the windmill rushed back and tore it down, lest the abbot's carpenters discover that it had been built out of lumber taken from land that belonged to the church.

The proprietors of land where there were no rivers to run mills welcomed windmill builders. The new machines saved them the expense of keeping animals to turn millstones, and enabled them to collect a fee for the use of the wind that blew across their land. Although the Bishop of Utrecht ruled in 1391 that no one could own the wind because no one could control it, landlords continued to charge millers for wind in some communities until the time of the French Revolution.

One day when the wind was not blowing, the Baron of Wassemer tried to collect his fee for "wind rights," only to hear the miller he approached demand that he produce some wind. The astonished nobleman protested that he could not possibly do that, and the miller replied, "But, sir, that's what I have been paying you for." To appease such tenants at least partially, many lords of the earth agreed to forbid the construction of any buildings near a windmill that might obstruct the wind. Laws restricting the height of trees and buildings near drainage mills are still enforced in Holland.

When personal rights to the wind's energy finally were no longer recognized, governments began to impose taxes on millers, the levies to be determined by the quantity of flour produced. The millers promptly found ways to evade them. Some operators devised ways of writing numbers in their account books that misled the tax collectors, and hid their revenues in secret compartments in their mills, the way some politicians have hidden cash in safe deposit boxes more recently.

The increasing population's demand for bread made milling more and more lucrative, but technological improvements also increased the cost of constructing a windmill. Builders began to lease mills to operators, the way modern computer manufacturers lease their products to merchants, governments, and other entrepreneurs.

In their day the windmills built in Europe several centuries ago were extremely modern devices, but few of us would now consider them significant sources of energy. At best they yielded only a few horsepower, but people appreciated that help then, because it relieved them of the labor of producing flour from grain by hand in their homes. Although those machines made cities noisier, more people enjoyed hearing them than objected to them, and the wind went blithely on its way as pure as ever after striking a windmill's sails.

It is fascinating now to trace the improvements that the Dutch, the English and their neighbors made in big vertical wind wheels, and even more interesting when one knows a bit about the horizontal rotors that the wind had spun for several centuries before.

## III

# *The Merry-Go-Rounds*

To imagine that things in this life are always to remain as they are is to indulge in an idle dream. It would appear, rather, that everything moves in circles.

MIGUEL DE CERVANTES

THE WORLD'S oldest windmills are in Afghanistan, a country about the size of Texas, squeezed between Iran, Pakistan, Russia, and China. It has no seaport, railroad, or navigable rivers, but its mountains are so high that geographers have called Afghanistan the Roof of the World. Up there the wind blows every day for nearly four months in the summer, sometimes up to seventy miles per hour. The Persians built their first windmills in deserts thereabouts before 1000 A.D. You can fly from Paris to the city of Herat now, but you would need a camel to visit what is left of many of the oldest windmills—and they might look as if they had been built upside down.

Explorers have found that the sails on the first windmills were under rather than above the millstones they turned, and that they were shielded by walls. The wind rushed in through slots in the walls to whirl the sails like revolving doors. Mills were built this way because millstones had to be turned horizontally. Animals and slaves had previously turned the stones by plodding around them in circles on the ground. Waterwheels under the stones had later turned them the same way, and no gears were needed when the sails were attached to a vertical pole that ran straight up to a millstone.

## The Merry-Go-Rounds

In time, the priests who climbed to balconies on high minarets to summon Moslems to their prayers must have noticed that the wind was usually stronger and steadier near the top of the tower than at the base. So millwrights soon began to place sails in the upper part of their structures to draw more power from the wind. In Afghanistan some of the older types of windmills were reported to be still running in 1952. But by the 1960s, when Hans Wulff wrote *Traditional Crafts of Persia*, all of the old windmills that he found being used had sails above the millstones. This was a significant improvement, and the mills he saw had been built to withstand whatever storms might come. In one small town on the fringe of the desert, Wulff saw fifty of them grinding six thousand tons of wheat seasonally with power from horizontal wheels, and was told that a diesel mill nearby could handle only about half a ton of wheat a day.

In a letter to *The New York Times*, June 3, 1973, Richard P. Leavitt described one of the ancient mills that he had seen near Herat. The millstones were in a one-story mud-brick building. Two slab-like walls rose from its roof. An opening between them let the prevailing wind rush in and directed it toward cloth sails. They were hung between arms attached at right angles to a vertical pole. A framework above them held the timber armature upright, and it was connected through a clutch to a millstone below the sails. This was a very sophisticated machine in its day.

Windmills were built the same way in China. There the sails were hung on light bamboo frames and used to pump sea water into basins so that salt could be removed from it. Some Chinese horizontal rotors have been reported still in use in this century. Neither China nor Afghanistan is likely to live in the past much longer, however, and historians of milling have planned long trips to study the primitive machines in these countries. Nearly everywhere else vertical fans have replaced horizontal sails.

Some modern engineers continue to favor sails mounted to be rotated horizontally on a vertical axle, because they will respond to a wind from any side and can run any kind of machinery on the ground directly below them. In the early 1800s there was a windmill with many such sails on the roof of a building at the foot of West Houston Street on the Hudson River front in Manhattan. We can see the machine in a colored drawing by J. W. Hill in the Museum of the City of New York. Another old drawing, recently found in Copenhagen, shows a tall tower with sails around a vertical axle, and horses inside the tower hitched to be driven around that axle to keep it turning when there was no wind.

In England in the nineteenth century, people marveled at three big horizontal windmills that Captain Stephen Hooper built. He made one of them out of a packing box that a Russian czar had left in London. When Czar Alexander I visited the city in 1814 he fell in love with a tiny church in Battersea. It was so beautiful that he decided to take it home with him. Assuming, as monarchs often do, that anything can be done with money, he had a packing box built alongside the church and planned to have the whole edifice moved into it on rollers. The people who worshipped there objected so fiercely to parting with the church that the czar had to abandon the project. So he went back to Russia without his packing box. Captain Hooper saw it standing beside the church, bought it, and sliced the sides of that big box into parts for a windmill.

## WIND-DRIVEN CAROUSELS

Merry-go-rounds for children to ride appeared first in French amusement parks about two hundred years ago. In the 1870s at the Jardin des Infants in St. Malo sails caught the wind to blow a small carousel around. The circular tracks were seldom more than fifty feet in diameter. At the Paris Exposition in 1867 a windmill on a track that wide, called "Moerath's wind wheel," was one of the most popular attractions. The builder had placed 31 big shutters around the wheel's circular track. They could be opened to expose the sails to a good breeze from any direction and closed to protect them from the wind when it was too savage.

Several farmers in Nebraska put this same idea to good use at the turn of the century. Tall, handsome, Professor Erwin Hinckley Barbour arrived at the University of Nebraska then, and toured the state by train where possible and horse and buggy elsewhere. Barbour became world famous by discovering the skeletons of a rhinoceros and other prehistoric creatures under Nebraska's bare sandhills. He won

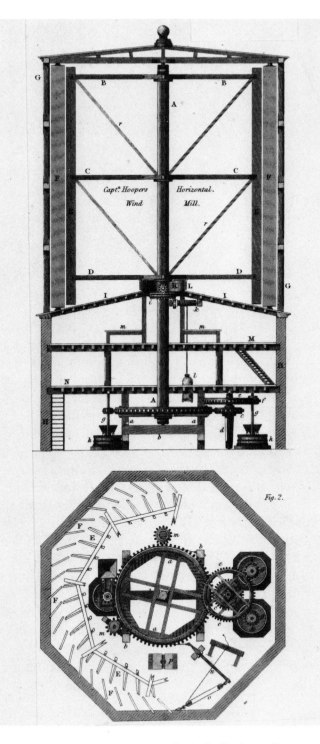

Drawings of Stephen Hooper's horizontal windmill, from diagrams published in 1816.

additional renown locally by his interest in homemade windmills.

On a ridge near Barbour's home in Lincoln, S. S. Videtto irrigated his garden with a homemade machine that looked so much like a carousel that small boys were tempted to jump on it. The wind drove sails 12 feet high around a circular track about 40 feet wide to run a pump for Videtto. This was a solidly braced, well-designed machine that Barbour hoped could be further improved.

The wind was often steady and strong in the smooth Platte River valley, and the water table was high—well-diggers often struck water within twelve feet of the surface. Professor Barbour thought windmill pumps made an ideal way to use the region's natural resources. He sent his students up and down the valley to observe, sketch, and photograph homemade machines running pumps. They found an amazing number and variety, including several that revolved horizontally and were called merry-go-rounds.

One distinctive type looked less like a merry-go-round than like a round animal cage mounted on top of a square wooden barn. This was because the horizontal wind wheel was surrounded by shutters similar to those around Moerath's wind wheel in Paris but much smaller. Henry Joenck had one of those windmills with a rotor about ten feet wide on his truck farm near Grand Island. It had cost him only about a hundred dollars, and had pumped water for his garden faithfully for fourteen years.

Charles Hunt manufactured windmills of this horizontal type briefly in Wichita, Kansas. He produced iron wheels from 4½ to 15 feet wide that could be mounted on low towers or barn roofs, and called them "Little Giants." But Professor Barbour in Nebraska was more interested in machines that farmers could build for themselves from scratch.

Near Berwyn in Custer County, W. Winn had constructed a machine that the professor especially admired. It resembled a shallow water tank on a tower. The horizontal rotor was inside a semicircular hood that revolved around the same axis. That hood shielded the sails on one side of the wheel while exposing those on the other side of the axle to the wind. A rudder attached to the hood turned it when the wind veered to keep the opening on the side from which the wind came. Professor Barbour thought that this contrivance was "well adapted to go on the tops of sheds, cribs, and outbuildings, thus doing away with towers," and could be simply and cheaply built.

He doubted, however, that the hood had to be a full half circle in width. So he designed an even simpler machine. It had a six-bladed horizontal rotor 20–25 feet in diameter, close to the ground like Videtto's merry-go-round. Barbour put four posts around this wheel for wooden gates that could be swung open or closed so that half of the sails would be fully exposed to the wind at the same time. Barbour wanted the state university to test and recommend his invention to home handymen, but it was forgotten.

Near the Colorado line, he pointed out, E. E. Blackman was successfully irrigating ten semiarid acres with a similar machine 24 feet wide. Its sails, he wrote, were like doors made of light wood, four feet wide and six feet high, and they were so attached to the driving shaft that they could "swing like a flag, edge to the wind when traveling against it, but broadside when traveling with it." Building that merry-go-round, moreover, had cost Blackman only the surprisingly small sum of $4.65 exclusive of labor.

"However impracticable and useless many inventions of our farmers may appear to be at times," Professor Barbour went on to say in a scholarly report on his study of windmills, "it is apparent that this inventive movement is an attempt to meet a demand, and who dares to say that the sum total of these inventions may not yet lead to the solution of some of our irrigation problems?" (More about other windmills that Professor Barbour studied will be said in Chapter XVII.)

Those problems were later reduced by building reservoirs and generating electrical energy with water turbines, but not until after some men had gone into the business of manufacturing horizontal windmills. Warren D. Parson claimed that the "Colorado wind engine" he patented and opened a shop to manufacture in Oakland, California, was the most powerful windmill then being built anywhere in the world. His machine had a horizontal rotor with a big weather vane above it. It could be mounted on top of a round railroad water tank or set up in the center of a city block to pump water for several homes.

29 ह‌

## THE "S" AND OTHER WIND TRAPS

American farmers and businessmen were less concerned about the efficiency with which the wind's energy was used than Commander S. J. Savonius in Finland. He experimented with cups for sails at Helsinki in the 1920s, and tested dozens of shapes in wind tunnels and open fields. He found then that if two half cylinders facing opposite directions were properly mounted on opposite sides of an upright spindle, the air flowing into one of them could spill into the other one and thus make the spindle whirl faster. Savonius therefore recommended horizontal sails shaped like the letter S, formed when two half cylinders were mounted as above. He thought that very powerful windmills might be built this way.

Savonius' invention was first used mainly for advertising. A single sheet of thin metal bent into the shape of a capital S would whirl around a vertical axle whenever a light breeze struck it. A merchant could have an ad painted on that metal, and S-shaped signs soon caught passersby's attention in front of many shops. Today, other uses have been found for Savonius rotors. They keep outlets for ventilating systems pointed the right way on the roofs of some industrial buildings, and are used in some of our finest meteorological and other instruments.

By telepathy if not otherwise, the Finnish commander's idea reached a home handyman in North Dakota whose name has been lost. The ingenious machine that he built in 1936 is shown in two photographs that Prof. W. J. Promersberger, chairman of the agricultural engineering department at the state university in Fargo, recently found in an old picture file.

To make an S-rotor for a windmill, the builder had cut two oil drums in half lengthwise. He put four of those half cylinders on the far ends of crossbars from the hub of an old auto wheel. He then perched that rotor on top of a plain wooden tower about twenty feet high. To connect the rotor to a pump, the builder used the framework from an old mowing machine. Every part of his windmill might have come from a scrap pile, but it must have worked, for one of the pictures shows a flock of sheep gathered around the tower to drink the water that the wind pumped for them.

Both the ideas of using old oil drums and of placing more than

*Homemade windmill with cut-up oil drums for rotor blades.*

two half cylinders around a single vertical axle may still be helpful in many places. Experimenters in several of our states have built windmills cheaply this way. And in Helena, Montana, Menasco-McGuinn and Associates recently proposed to manufacture wind-driven turbines with horizontal rotors to generate electricity for a solar heating plant they were working on.

Most engineers in this century, however, have found the propellors of aircraft more fascinating than sails made out of oil drums, and have tried harder to perfect airfoils for windmills. In France in 1929, Monsignor Georges Jean Darrieus substituted two long blades like those of a propellor for the customary sails on a big windmill. The wind spun his rotor delightfully fast and electricity was generated by it at Le Bourget airdrome. At Cocoa, Florida, Jesse C. Donaldson tried similar blades on a horizontal rotor without attracting much attention. Other more recent experiments will be cited later (in Chapter XV).

31 ❧

In the 1850s a German physicist, H. G. Magnus, studied the sidewise thrust of air on a smooth ball or cylinder that is rotating while moving through it. The "Magnus effect" was used later to propel a ship across the Atlantic. This tempted New Jersey public utility companies to see if that sidewise force could be used to generate electrical current economically. *Popular Mechanics* reported in 1932 that those companies proposed to put twenty tall cylinders on big railroad flatcars on a circular track. Each car in that strange train was expected to cost $100,000, and the plant was designed to generate 40,000 kilowatts of power. But the utility companies abandoned the project after building and testing one big cylinder on a stationary platform at Burlington, New Jersey.

Several engineers in this century have thought of concentrating the wind's energy by catching it in the wide ends of big funnels. The jets at the opposite ends of those horns might then be used like streams of water flowing down from reservoirs to turn turbines. A French physicist, Bernard Dubois, suggested that warm air flowing up a chimney might be concentrated similarly to generate electricity. Both Dubois and a fellow countryman, Georges Claude, submitted plans to do this to the French Academy of Sciences. That body of cautious advisers to government was so favorably impressed by Dubois' proposal that they recommended it be tried in French North Africa, which the Atlas mountains and scarcity of fuel in French Algeria made a natural site for such research.

Some of the Atlas mountains rise sharply for about 6,600 feet from a hot desert. On top of them the air is under less pressure, and is often at 30–40 degrees Fahrenheit, which is much cooler than at the base. Professor Dubois proposed to take advantage of that difference in the pressure and temperature. He wanted a glass roof built over a concrete apron on the desert sand at the foot of a mountain. Warm desert air would be forced into that hot place, and from it the hot air would race up the side of the mountain, via a porous cement chimney, to turn a turbine on the summit. The chimney conceived would be about 30 feet wide, and the air might zoom up it at 180 feet per second. But building a big plant like that to test the professor's plan cost more than the thrifty French were then willing to invest in Algeria, and Dubois' plan was shelved.

Another scheme for a chimney power plant came to French researchers in an aerodynamical institute at St. Cyr, and was based on a different scientific principle. Willy Ley, in his book *Engineers' Dreams,* describes this plan vividly: "The device consists of a sheet-metal chimney just strong enough to stand up under the wind, which is open at the top and has a number of large holes around the bottom so that the air can enter freely. The height of such a chimney would be about 60 or 75 feet. At its crown a 'collar' is attached, a fairly short section of a similar chimney of double the diameter. This collar is held in place by narrow struts, so that air can enter into the space between chimney and collar without any obstruction."

The air entering the chimney through the holes near the bottom would be moved by air blown into the space between the collar and the chimney by a wind from any direction. The only outlet would be the open top of the chimney, and the air would have to turn a turbine inside the chimney above the air holes to escape.

No government or utility company has yet built such a trap for wind from any direction to enter. The idea that the French theorists wanted to try, however, did not die. Some engineers still hope to demonstrate a way in which wind can be given more momentum after it has been caught.

## SURVIVORS AND DESCENDANTS

In the 1940s Greville Bathe, a modelmaker in Pennsylvania, wrote a fascinating book about horizontal windmills. He illustrated it with drawings and photographs of several of the machines mentioned here and many others. By then hundreds of patents had been obtained on horizontal windmills, but few of them ever turned out to be profitable to the inventors, and examples of horizontal machines became rare sights.

The only one recently reported to be still running is at Winnemucca near Desert Valley amidst the mountains in northern Nevada. C. L. Swet built this machine in 1958 and has used it ever since. It is a more modern version of a horizontal windmill that Swet's father built in 1921. The outside vanes are stationary and direct the wind to the sails that they enclose. A hydraulic governor keeps the machine from running too fast. It has withstood gales up to 80 m.p.h.

With this turbine Swet has pumped water, charged storage batteries, turned a tumbler to polish rocks, and kept an electric sign lit for his roadside business. "The only attention given to it," he reported when *Popular Science* published a snapshot of it in 1973, "has been to grease the bearings twice since installation." Publicity put Swet to more trouble than he had ever had with his windmill. He was promptly swamped with requests for plans to duplicate it.

A much older horizontal windmill is still upright but in sad shape in northwestern New York. It is on a hillside near a fine old home on the Old Gladden Farm not far from Randolph. The windmill, built around 1890, looked like an oddly shaped barn about to tumble down when Eric DeLony, a young man employed by the National Park Service, discovered it in 1973. The foundation had been weakened by erosion of the hillside, and the old mill's sails had been removed to repair an outbuilding.

DeLony, an authority on historic buildings who has seen many of them, found this "the most incredible structure" that he had ever inspected. He learned that the Gladden windmill had once been used not only to grind grain into flour, but also to sharpen tools and crush apples for cider. The New York State Historical Society had proposed to move it to a Farmers' Museum in Cooperstown, New York, but found the cost prohibitive. DeLony promptly arranged for engineering drawings to be made of the parts of the machine, most of which were still on the farm, so that it could be restored or reconstructed.

More experimental windmills are being built in the United States now than were in the 1940s, and some of them have horizontal rotors. Although usually made out of new materials and designed more carefully than those constructed long ago, many of them are based on the same principles as ancient machines. Wholly new ideas rarely occur to most of us. When we confront new problems, we usually fall back on old ideas to solve them. Most inventions are merely modifications for new purposes of previous inventions. When interest in wind energy started to increase, engineers came up with new kinds of power plants that are reminiscent of much older ones.

The first air-flow engines, built centuries ago in deserts like those in Afghanistan, had rotors that revolved in the plane in which the air most obviously traveled. But the machines built later with big sails nearly at a right angle to the ground were more efficient than the Persians' primitive merry-go-rounds, and more beautiful. People preferred them nearly everywhere.

IV

# The Mills on Posts

The wind blows where it wills, and you can hear the sound of it,
but you do not know whence it comes or whither it goes . . .

ST. JOHN, 3:8

W H E N   L E O N A R D O   da Vinci around 1500 drew a
sketch of a windmill with long sails for the wind to whirl vertically,
it was difficult to construct a strong tower high enough for them.
The picture in Leonardo's notebook showed four big sails attached to
an axle on one side of a round tower, and many Europeans must have
doubted whether such sails could be kept facing the wind where they
lived enough of the time to be worth building. But seamen's experience, craftsmen's skills, and women's fatigue led someone to devise a
solution for the problem. Artists have admired post mills for centuries.

This solution to the problem was remarkably simple: the whole
machine was perched on a post so that the miller could turn it to
face whatever direction he chose. Where and when this first was done
is uncertain. Although millstones weighed as much as an auto may
today, mills to house them were seldom much bigger than our single-
car garages. Nevertheless, balancing a millhouse, in addition to several sails, on a single post took both skill and strength.

The first windmills built in our own and many other countries
were constructed this way. Unfortunately post mills were so vulnerable to storms, lightning, fires, termites, decay, and neglect that most
of those built on our side of the Atlantic lasted only about twenty
years. Mark Twain never saw a post mill until he went to the Azores.
He thought it was a ridiculous way to build a mill. "When the wind

35 ஓ

changes," he wrote, "they hitch on some donkeys and actually turn the whole upper half of the mill around until the sails are in the proper position instead of fixing the concern so that the sails could be moved instead of the mill."

The lower "half" of that mill was simply a strong post braced to support the housing for the millstones. Elsewhere the supporting post was sometimes sunk in a mound of earth or concealed by walls around it, but often it was left fully exposed. No two post mills were exactly alike; each one was built by hand for a particular person at a specific place with whatever material the millwright could find.

The builder of a seventeenth century post mill that was recently restored in England had chosen "the heaviest piece of oak in Hertfordshire" for its mainpost. "Do not take a tree from a wood on poor soil," one millwright warned others, "and beware of black Italian oak which is heavy and not always suitable. Choose a good English tree growing in a hedgerow on strong heavy stone soil." Oak from trees stripped of bark in the spring and not felled until autumn was said to be less liable to split or be weakened by worms.

The mainposts were usually taller in the northern countries than in southern lands. To turn a millhouse on a high post in the desired direction, a tailpole was attached to the floor of the housing at the rear. It curved down to the ground so that a man or beast on foot could pull or push the thing whichever way was necessary to keep the sails facing a favorable wind, or out of the way of a wind so fierce it might destroy them. In Spain and Romania, the millhouses were sometimes so close to the ground that the millers could push them around with a crowbar.

A distinctive type called the "caviar" mill was once common in the Loire Valley of France. Wind from the Bay of Biscay blew up that valley, and the river was a poor source of power because it dried up in the summer and overflowed in the winter. The cabins on the posts of caviar mills were nearly cubical and poised on the tips of broad conical bases. Grain from the plains of Anjou could be stored in those bases, and they also served as wine cellars for the vineyards on the riverbanks.

Especially on bare flat plains near the North Sea the post mills were ornamental as well as functional. Grandfathers could not build prettier dollhouses for little girls than the tiny millhouses. They were distinctively designed and decorated; millers kept them tidy and brightly painted, and gave them fanciful names. Their height made them helpful landmarks for travelers; cartographers drew little symbols to designate their location on maps, and those windmills also often served as beacons, observation posts, and semaphores.

Like sailboats, the European windmills were usually referred to as feminine, possibly in part because of their beauty and partly because of their coy behavior. Even their voices seemed almost human. They whimpered when the wind awakened them, rumbled when they went to work, and chattered when running at top speed. Fathers taught their sons the tricks of managing those mills, and in some families the miller's trade was passed on for several generations.

### WILLIAMSBURG'S WINDMILL

The only authentic reproduction of a post mill in the United States until very recently was one in the restored historic part of Williamsburg, the Colonial capital of Virginia. You can see it from a highway on your way to Jamestown; it is well worth stopping to study.

William Robertson, who was clerk of the British governor's council for thirty years, built a post mill in Williamsburg about 1720 near the palace where the governor lived. Robertson's mill was one of two that the best cooks in Williamsburg visited every week or so. The corn meal the wind ground there was hard to keep, and was best when fresh. Seeing the wind freely spin the big sails of those wooden engines on the way to and from work may also have nourished the confidence of that little town's leading citizens that Americans could successfully defy a distant British king.

Unfortunately no remnants of Robertson's windmill could be found early in this century when the Colonial Williamsburg Foundation set out to restore the oldest part of the town. The director of architectural research, Paul E. Buchanan, began with only meager clues to the windmill's location and appearance, but had the best scholarly advisers obtainable. M.I.T. professors helped decide, for instance, how wooden teeth for gears were made and shaped 250 years ago. Rex Wailes, the noted British authority on early windmills, came to Williamsburg to answer questions about other details. Possibly the oldest of several famous mills still standing in England was built in the 1600s at Bourn, and the rebuilders of our colonial town decided that Robertson's mill probably had closely resembled it.

A tiny working model of Robertson's mill was made then by Edward P. Hamilton, a retired investment counselor and former director of Fort Ticonderoga. Full-scale parts were produced from that model (which Buchanan kept in his office) and tested in an airplane hangar. After five years of research, nine more months were spent reconstructing the Robertson windmill. Oak from old trees in West Virginia was used, and the millstones were cut from quartz-bearing granite found at Mount Airy, North Carolina.

A black workman who helped to reconstruct the post mill became so interested in it that he learned how such mills were managed, and stayed on to be the miller until his death several years ago. His son, whom he had taught, became the miller then. When I interviewed him, Cornelius Black was wearing a yellow colonial jacket and knee breeches and supervising the operation of the windmill from an office in a shed nearby.

A split rail fence encloses the town lot where the reconstructed windmill stands. The long rectangular sails sweep close enough to the ground to knock a man down, and soar high above the ridge of the peaked roof of the white frame millhouse pivoted on a post. A tiny weather vane on the ridge, with W. R. for William Robertson painted on it, points the way the wind is blowing to help Black keep the sails in or out of its path.

The millhouse is only about 10½ feet wide and 11 feet long, but is raised 11 feet above the ground on its post. To enter it, you must climb a flight of stairs to a balcony at the rear. A heavy pole slopes from the millhouse floor through the stairway to a wagon wheel on the ground. Black can turn the building in any direction by lifting the flight of steps a few inches off the ground with a lever and pushing the whole mill around with the wagon wheel on a circular track 68 feet in diameter. When lowered again, the wooden stairway braces and steadies the millhouse.

The sails of the Williamsburg mill are long strips of natural Scottish linen, tied at each end to a flat pine framework on each side of four Sitka spruce masts. To furl the four sails, the mill must be stopped four times. The operator can apply a brake to do this. Then he unties each sail from the far end of its framework and twists it into a tight cord nearly to the hub. Each sail is about 25 feet long.

The axle that those sails turn protrudes from the slightly curved front of the millhouse and is tilted skyward about ten degrees. This tilt and the sloping stairway at the rear make the building look at first as though it sloped from front to rear, though its floor is actually level. Two pairs of windows on opposite sides of the house indicate that there are two floors in it.

The whole mill weighs about seven tons, and the mainpost that supports it is two feet square. It was hewn by hand from an oak tree that was a sapling in 1675, and is braced by foot-thick timbers that rest on four brick pillars. The mainpost extends from the ground to the second floor of the mill. There it is capped by a round wrought-iron bearing on which a horizontal timber is centered. This timber, called the crowntree, is the central timber of the millhouse. It bears the weight of the stones, the housing around them, and every working part of the mill.

The first bedstone used weighed 1,700 pounds, and the first runner 2,500 pounds. It took a 5-m.p.h. wind to turn that runner, so lighter stones were cut to replace the first pair. Now the wind will turn the runner when it is blowing only 3 m.p.h. With this pair Cornelius Black can produce about 180 pounds of corn meal an hour, and he has run the mill in gales up to 60 m.p.h.

Hundreds of visitors pause on trips through Williamsburg nearly every day to peek into the windmill. Cornelius Black often needs two helpers to attend to the mill properly and answer questions.

From the mill's balcony you can look in and see both what the miller does and how everything works. The mill's speed can be reduced by tightening a bent hickory band around the rim of a brake wheel more than seven feet wide on the slanting tapered shaft that the wind revolves. If someone were to apply that brake too hard, however, the friction of the wheel's turning under the brake might cause the wood to ignite, setting the whole mill afire.

Fifty-one hickory pegs driven into the face of that brake wheel around its outer edge make it also a driving gear. Those teeth fit into the slots between dowels of a "birdcage" gear to spin the vertical shaft that whirls the upper millstone. Every man who remembers how hard it was to make a model windmill with wooden Tinker-toy parts when he was a boy can appreciate how much thought and work went into making those gears.

The millstones as well as the big gear are on the upper floor of the mill. The miller hoists corn up to them through a trapdoor and stays below while the machinery is running. The runner turns on the point of a spindle above the bedstone, and corn flows into the eye or center-hole of the runner from a hopper vibrated by an attachment to the wind-driven shaft.

When ground, the meal is thrown out into a wooden cabinet around the stones. From there it flows down a canvas trough to a sifter on the lower floor. Cornelius Black stands beside that sifter and feels the meal with his hand, the way millers did centuries ago, to make sure that it is being ground to the proper fineness. He can control the fineness by varying the space between the stones and the rate at which the grain flows from the hopper. Such adjustments are made with ropes and wooden levers.

From his station beside the sifter, the miller can keep an eye on the machinery above him through the trapdoor, and on the weather outside through a window nearby. It is difficult to imagine, when you see all this, how such an intricate machine could have been constructed long ago in a newly discovered land without the help of modern tools, cranes, and big trucks. Even in our century, it was a challenging and time-consuming feat.

Williamsburg's reconstruction of Robertson's mill will almost certainly outlast the original mill because of unusual protective features that its architect proudly points out. To shield it from lightning, wires were run down each mast of the rotor and the mainpost to the ground. When a hurricane warning is sounded, the miller can lift the whole structure off the mainpost. He does this by erecting light aluminum rods at each of the four corners of the millhouse and extending them with turnbuckles to support the house firmly on four legs rather than one. Storms blew down many of the first post mills before they earned enough for the builders to repay them for their work.

## THE MILLER'S TRADE

After seeing how Williamsburg's mill is run, many persons go on to a restored colonial bakery behind the Raleigh Tavern where

*Cut-away view of post windmill model.*

they can buy a bit of "the finest meal ground in Williamsburg," fresh from between the millstones casually turned by the wind.

Hush puppies, spoon bread, shortening bread, and other regional treats were first made from meal ground this way. It often was heated less than meal ground between rollers, and some gourmets believe bread baked from it tastes better than the stuff we now find in supermarkets. Others deny this. Even a veteran Dutch miller insists that it is "just a fable" that the corn meal ground between stones in his windmill tastes any better than if it were ground by the rollers in a modern Minneapolis mill.

A children's encyclopedia tersely explains that "old process meal is better flavored and more nutritious, but new process meal keeps better." Producing a hundred pounds of meal the old way took only about two bushels of corn. Three bushels may be needed the newer way, yet this may be worth while. Modern methods are certainly more sanitary, and most of us live longer than our forefathers.

Men went on building post mills for scores of years mainly because they could be run in places where water wheels were useless. A post mill was small enough to be moved to a windier site if the first place it was erected was not satisfactory. Dozens of oxen sometimes were hitched together to tug a mill nearly intact for a short distance. Mills were moved so often and so far that some builders numbered each part to make it easier to reassemble the pieces properly. In the Netherlands the first choice of a site was often an old mound or powder magazine because a high point was nearly always windier than a low one. The mainpost and bracing around it were likely to be left open at first nearly everywhere, but could be enclosed by a "roundhouse" later if the site turned out to have been a wise choice.

The miller stored grain in the roundhouse under his mill, and kept his tools there, possibly including fishing gear to use in his spare time. Not that he had much. There was always something for him to do. Periodically the millstones had to be separated and the furrows in their faces sharpened. This was done with two chisels on the end of a hammer-shaped "millbill." When the stones were properly dressed the miller had to resharpen the cutting edges of his millbill.

The wind's whimsicality made running a post mill more demanding than managing a water mill. Sudden fierce gusts "tail-winded" many millhouses by blowing them off their mainposts. This could happen at any time of day or night. Ministers are even said to have interrupted church services at times to tell a devout miller to go tend his mill.

A single mill might serve a thousand persons, and its operator was a middleman between them and nature, a private citizen performing a public service. He judged both the quality of the grain that farmers brought to him and the meal that his stones ground. Neighbors sometimes challenged his decisions, and Chaucer's description of a miller was often recalled:

> His was a master hand at stealing grain.
> He felt it with his thumb and thus he knew
> Its quality and took three times his due.

Laws to "rectifie the great abuse of millers" by limiting their fees were enacted centuries ago nearly everywhere. Other laws forbade a miller to keep chickens or pigs, lest he feed them grain that was left in his custody. In Virginia, indentured servants or slaves often ran mills for wealthy owners, whose greed may have sometimes encouraged rascality.

Observers hovered around windmills like hawks, and questioned the millers the way tourists at Williamsburg question Cornelius Black. Hoisting and lowering heavy sacks, racing up and down steep steps to keep sails set right, seeing that the runner stone was turning the way it should, dickering with customers—all this could exhaust even a hardy man. Yet many millers became popular dispensers of news and gossip, and highly regarded weather prophets, denounced when they erred like today's radio and television seers.

To thrive, a miller had to be not only a competent mechanic and handyman but also a tactful expert on crops, cooking, and community affairs. There were no schools where all the arts a miller had to practice were taught. Despite the trade's many hazards, millers frequently became influential citizens, political bosses, and even bankers. People addressed them as "masters" for the same reason that we now call learned men and women "doctors" regardless of their occupations. Like high officials, schoolteachers, and jailkeepers, the millers then were often excused from military service. Until a couple of centuries ago the ministry was the most prestigious profession that a man might choose, and in some countries running a flour mill was considered the next most honorable calling.

### EARLY AMERICAN LANDMARKS

Post mills were prominent landmarks in several of the American colonies. Sir Henry Clinton, who commanded British forces sent to quell the American Revolution, carried a map of Virginia with big symbols for post mills on it at three key points.

Sir George Yardley is credited with building our country's first post mill, in about 1620. Yardley had fought England's foes in Holland and was shipwrecked in Bermuda before he arrived in Virginia in 1610. Yardley became Virginia's colonial governor in 1618 and soon built a post mill on the thousand-acre plantation that was granted to him.

Yardley's land was about twenty miles from Jamestown on a tip of land in the James River. He called his plantation Flowerdew Hundred, because his wife's name had been Temperance Flowerdew. There were then nearly fifteen hundred colonists to be fed, and the approximate site of the Governor's post mill has been known as Windmill Point since 1670. Mr. and Mrs. David A. Henderson, who now own the Flowerdew plantation, established a foundation a few years ago to support archeological research and reconstruction of some of the early structures on it including Governor Yardley's post mill. Professor Norman Barka, head of the Department of Anthropology at the College of William and Mary in Williamsburg, has directed much of this work. Mrs. Barka has helped in the work by writing a scholarly report on the Flowerdew post mill.

From meager scattered clues a model of Yardley's machine was built and carefully studied. One of England's foremost millwrights, Derek Ogden, then was authorized to construct full-sized parts with fine, seasoned oak for shipment to Virginia. At this writing, the reconstruction of the Yardley mill is scheduled to be completed by 1977.

Windmills were also built elsewhere in the old Middle Colonies. In Chesapeake Bay a British Admiralty court note for 1635 read: "Our principall imployments for our men were in making two windmills." Just where these windmills were built is not known. Another windmill in the Chesapeake Bay area, however, is known to have served a self-sufficient community started in the 1800s on a plantation originally granted to a settler by Lord Baltimore.

For this settlement John Anthony LeCompte Radcliffe had a post mill built that ran until the great blizzard of 1888 destroyed it. His youngest son, George L. Radcliffe, whom Maryland later sent to the United States Senate, never forgot that windmill. On his ninety-fifth birthday, a year before his death in 1974, the senator's friends and neighbors dedicated a reproduction of his father's post mill to him.

The millstones and the stairway of the old Radcliffe mill were the only original parts that James B. Richardson, a local boatbuilder, could find when he set out to rebuild this mill. His son-in-law, Thomas Howell, Jr., helped to work out the plans for it, and they solved many of their problems by trial and error, like the early mill-wrights. The mainpost was cut by hand from a white oak that grew in the nearby Spocott woods.

The reconstructed Radcliffe mill, which differs from the Williamsburg post mill in several respects, was reported to be able to develop up to fifty horsepower. It is six miles west of Cambridge, Maryland, within sight of a state road, and only a short drive from Washington, D.C.

Artists have done more than social historians and reconstructors to keep us from ever completely forgetting windmills. Rembrandt was one of the first of many famous painters who have portrayed them. His somber painting of a post mill at dusk is one of the most beautiful in our National Gallery of Art in Washington. The mill depicted stood on a high bluff overlooking a river. In the foreground you can see women washing clothes in the stream and a man who has lowered the sails on his boat to use his oars. This, a guidebook to the gallery says, "is one of the most profound and moving pictures ever painted . . . so evocative that it seems to have endless depths of romantic meaning."

Our National Gallery also has an etching by Rembrandt of a windmill that belonged to one of his grandparents. It is less alluring than the post mill that he depicted in oil, it looks more decrepit, and its sides sweep upward to a cap rather than a housing for millstones. This was a smock mill, easier to erect and run than a post mill. By the sixteenth century smock mills had begun to replace post mills in Holland. Both types were brought to the New World by Dutch and English colonists.

# The Dutch Windmills

They were our first machines and they opened a new era.

JOHAN VAN VEEN, MASTER OF THE FLOODS

JAN ADRIAENSZOON was the first inventor of a windmill about whom much is known. He was born in northern Holland in 1575. Admirers later called him *Leeghwater*, which means "empty water," because he showed the Dutch how to enlarge their land by draining lakes. The Persians had used the wind to water their gardens, and Leeghwater used it to remove water from flooded fields.

In Leeghwater's time, Holland was a miserable place to live in. Ice Age glaciers had left a maze of mudholes and river mouths behind them when they retreated into the North Sea. Tides swept in to flood the bare lowland. The few Frisians who dwelt there in the first century A.D. lived on mounds surrounded by water. Pliny reported that they survived by "burning mud, dug with their hands out of the earth and dried to some extent, more in the wind than in the sun." When more poor people with no better place to go moved into that "nether world" they stripped more peat off the land to warm their huts, and water spread over more fields.

Post mills had been built in the Netherlands centuries before Leeghwater was born, and farmers already were using windmills to move water from one field to another. Moving water from field to field did not dispose of it, however. Even large settlements such as Amsterdam were imperilled when water crashed through walls and dikes. Ten thousand persons drowned in a flood that wiped out sixty-five hamlets on a single November night in 1421. An effort on a larger scale was needed to reclaim land permanently.

The Dutch were equal to the effort. They didn't lack for ingenuity, energy, courage or determination. Dutch dikemasters discovered streamlining principles that engineers still respect, and their mill-wrights produced crude airfoils. This was done when they found that the "pull" of sails on a shaft was greater if the sails were put behind rather than centered on the masts or "whips" that supported them. A board could then be placed on the other side of the whip to give it a blunter edge, making it a more effective airfoil.

The Dutch already had several kinds of windmills. For flour mills on posts they had fetched oak and other lumber from other countries. They brought pitch pine for long masts, for instance, from the forests of the Scandinavian peninsula, and boxwood from the Pyrenees for sturdy small parts.

Post mills like those designed originally to turn millstones could be used to lift water if they were constructed with hollow mainposts to accommodate a rotating shaft. Hollow post mills were called *wipmolens*, and were used to turn big wooden wheels with scoops around them for lifting water. But those buckets could only lift the water a few feet at a time, and a *wipmolen* had to be over or along-side the pond to be emptied. Such a mill might have to be moved several times to drain a large field.

The problem of mobility was solved with another kind of windmill called a *tjasker*. Whoever introduced it must have remembered or rediscovered how the Greeks had used a long auger rotating inside a slanting tube to draw water up the tube. A *tjasker* was simply a wide pipe containing an Archimedian screw that could be turned by the wind striking sails on its upper end. When one end of the pipe was dipped into a pond, the sails on the upper end were elevated enough to intercept the wind. This device, which could be moved around easily, could raise water as much as ten feet at a time. But neither *tjaskers* nor *wipmolens* solved the problem of disposing of the water after it was lifted. That was where Leeghwater played his part.

Leeghwater had already invented an inexpensive little windmill for lifting water. Its wings were plain wooden boards parallel to the masts. The whole thing was easy to assemble and move from place to place, but it didn't solve the problem of getting rid of the water it lifted. For that problem, Leeghwater turned his attention to larger mills, including smock mills.

Smock mills were so called because their profiles resembled those of the artists' smocks that many workmen wore then. On those wind-mills only the cap on top of the tower had to be turned to keep the sails facing the right way, and this could be done either with tailpoles like those on post mills, or with gears inside the towers.

It was the smock mill that helped Leeghwater find a way to drain the land of water. The big tower mills that we call "Dutch" today developed from smock mills rather than from *wipmolens, tjaskers* or Leeghwater's first little waterlifters. Those giants provided the energy needed for industrialization of the Netherlands.

### HOW A LONG WAR BEGAN

While Leeghwater was developing his little water-lifters, the Dutch East Indies Company was becoming fabulously wealthy from trade between Europe and the East. Mynheer Dirck van Oss, one of the "seventeen gentlemen" who directed the Company, conceived a bolder new venture than any mighty American corporation has undertaken lately. The wind was propelling ships to and from distant lands for the "gentlemen," and Mynheer van Oss wondered if it might not also be used to increase his homeland's resources.

If there had been management consultants then, they surely would have advised the company that the available windmills were too small to defeat the savage sea. But there was then no Mumbel, Fumbel, Grumbel & Stumbel, Inc., to restrain rich men from acting on noble impulses. So the company hired Leeghwater to head the effort to reclaim land that the sea was keeping the Dutch from using.

Leeghwater decided to begin by draining the big Beemster Lake. It would require fleets of both large and small windmills, but might be done if the lake were first isolated from the rest of the country. To do this Leeghwater had two rings of dikes built around the lake with a canal between them. Then he used half of his big windmills to pump water out of the Beemster into that circular canal. From that reservoir other windmills then could lift the water over the outer ring of dikes, where it could be dumped into streams in which it could flow off to the sea.

The water in some parts of the Beemster had to be lifted more than twenty feet. The windmills had to work together like parts of a single complex machine. Every step had to be carefully planned and precisely timed. Nothing could be done when the wind was too weak or too strong, and several months passed before people could wade into the muddy bottom of the Beemster Lake to catch eels.

That reclaimed bottom land was divided into *polders*. These were fields separated by small canals inside the circular canal on the perimeter of the old lake. Little meadow mills kept a farmer's fields sufficiently dry by pumping water into those drainage canals. Cattle could drink from them and boats could enter and leave the inner canals through locks.

Before the expensive project was completed, the sea crashed through the dikes and filled the Beemster Lake again. Despite this and other setbacks, however, the desperate Dutch people continued to follow Leeghwater's system of dividing the drained land into *polders* and building dikes and canals. It required accurate surveying, well constructed dikes and canals, and scores of windmills. Dutchmen mastered the necessary arts and began to win their battle against the sea.

Leeghwater became an advisor to Prince Maurice on military as well as civil engineering, and before he died in 1650 he worked out a plan to remove water up to thirteen feet deep from an even larger area than the Beemster had covered, in a scheme that required 160 windmills. His account of how the job might be done evoked so much interest that it was reprinted seventeen times. This great project was not undertaken, however, until 1848, when the Dutch had steam engines as well as windmills to run pumps.

Descartes was said to have remarked that "God made the world but the Dutch made Holland," and proud Dutchmen often quoted the remark. They believed in both miracles and their own strength, not without justification. When their millers formed a guild they named it after Saint Victor, a Roman soldier who became a Christian before Emperor Constantine did. Victor smashed an altar to Jupiter and was sentenced to be crushed to death between millstones. But the millstones were reported to have crumbled when he was thrust between them, and the executioner had to draw a sword to kill him.

Things happened in Holland that seemed fully as miraculous. After the great flood in 1421 a baby in a cradle was washed safely onto a dike—thanks to a cat that had kept the little bed afloat on the waves by jumping from side to side in it. This occurred near Rotterdam in a region still called Kinderdijk (children's dike); more than two dozen *polder* mills have been preserved there and Dutchmen now run some of them on Saturdays for tourists' benefit.

## WINDMILLS AS HOMES

The Dutch drainage mills usually were in lonelier places than flour mills. Their operators had no streams of customers to dicker with, and so made their mills inviting shelters for travelers. In the marshes the sides of small windmills were sometimes thatched with straw, which made them look as though nature had produced them.

Even when the miller's home was no larger than a modern camper's trailer there were invariably two doors to it, on different sides so that a person could step in or out without being hit by a sail. The larger mills often had living rooms on the ground floor from which lamps shown at night through tiny window panes, and those homes were kept well stocked with food for guests.

Dutch windmills seldom had chimneys. The millwrights had thought to put the smoke from the hearth fire to work, rather than let it escape. Smoke was piped only into the upper part of the tower, where it helped preserve the wooden gears before filtering out through chinks in the siding or roof.

The four broad sails of the windmills quickly became not only symbols of comfort and power but also bulletin boards for news. Every Dutch child learned that the position in which the miller stopped the sails indicated the current state of affairs.

If one pair of sails pointed straight up, the miller expected to let them start turning again soon.

If both pairs were at sharp 45-degree angles to the ground, it meant the mill would be idle for quite awhile.

If the operator had stopped the sails from turning just before the uppermost one was vertical, it meant he was celebrating good news.

If he had stopped the top sail only a few degrees past the high point in the direction the wind was blowing, it meant that he was mourning.

In the cities the windmills were as tall as church steeples so that their sails could catch the wind above other buildings. One that can be seen from afar from all directions still stands in the old part of Leiden; floodlights are turned on it at dusk now, and it is decorated with flags on feast days. When German troops invaded The Netherlands in both of this century's world wars, windmills warned the people of the enemy's advance. Windmills also became targets for artillery shells and bombs that demolished many of them.

Proud windmill owners gave their big mills names as imaginative as any now put on American cars: The Seeker, The Cat, The Lion, The Stork, The Seagull, The Cornflower, The Prince's Garden. Symbols carved in wooden boards were placed on the lower edges of the caps behind the sails. Those embellishments were called "beards," and were copied in the coats of arms of many noted families.

### INDUSTRY'S BIG WHEELS

A Dutch smock or tower mill could be both profitable and spectacular three centuries ago. In a grain mill's wide high base there might be ample space both for a fine family home and for storage, a luxury few of us can afford today. The machinery could all be in a multi-storied wooden tower overhead.

The sails, revolving around a hub on the roof, would be too high to hit a pedestrian, or to be kept facing the wind with a simple tailpole. So in northern Holland, where the weather was inclement, a big mill might be made an "inside winder," with gears and wheels within the tower for redirecting the sails. But the miller still would have to step out on a balcony where the tower met the base to furl or unfurl the sails.

Elsewhere "outside winders" were often preferred because they were more beautiful and fun to watch. The balconies around tower mills were called stages, and the operators of outside winders used them to steer as well as adjust the sails. Every passerby could see and comment on the miller's performance of those duties.

Instead of a tailpole, a horizontal bar was attached to the rear of a high tower mill's cap. Poles then were run down from its ends and centers to a capstan wheel on the balcony. This triangular framework resembled the tailpieces that peasant women often sewed on the rear of their bonnets. It enabled the miller to turn his machine's big sails into or out of the wind's path with a wheel on the balcony.

This is the kind of Dutch windmill that is most admired today. Inside a tall tower there could be many floors. The top one directly below the cap could be used as a "dust floor" to catch the fine white powder that rose from the millstones. The next floor down could be a "bin floor" to which the grain brought to the mill was hoisted. From there it could flow down chutes alongside a whirling vertical shaft to a "stone floor," sometimes wide enough to accommodate two or three pairs of millstones. From that floor the ground grain could flow down to an even wider "meal floor" directly below the stones at the level of the balcony around the tower.

Dutch windmills ground lime, chalk, cocoa, mustard, and snuff as well as grain. Some ground oak for tanneries. The sails of others ran paper mills and fulling mills for the treatment of cloth.

The inventor of a wind-driven sawmill patented in 1592 called it The Damsel, and many descendants from it ran saws in the next century. These were called *paltrok* mills. Long sheds, like the wings of a ranchhouse, were attached to the lower part of a *paltrok*. Wooden cranes lifted logs from the water into those sheds, from which long logs were shoved through frames of saws.

The *paltroks* built in Holland's Zann region were remarkable structures somewhat similar to post mills. They were built around king-posts that rested on brick pillars in the centers of round brick walls. Most of a *paltrok's* weight was borne by the kingpost. But on the circular outer wall of the foundation there was an oaken sill called the "winding floor." On it, the mill could be pushed around on rollers to keep the sails pointed the right way.

While the *paltroks* filled the lumber yards with wood ready for carpenters, another kind of windmill produced oil, which it crushed from seeds in iron cylinders. The seeds were then heated over fireplaces and further pulverized with rams and stamps. The booming strokes of a big oil mill could deafen a workman in it, and keep others awake miles away. The simultaneous screeches from saws, the shouts of teamsters, and the howls of animals, added to those loud booms, made some Dutch towns the noisiest places then imaginable. But the same wind that provided the energy for new industries carried away the odors they produced.

The Zann River's banks were famous a couple of centuries ago for the wind-driven paper mills built there. One of them, called The Schoolmaster, has been well preserved. There you can also see an oil mill built in 1620 and completely restored in 1939. Another industrial mill nearby, called The Seeker, was originally a drainage mill used in Leeghwater's project to reclaim the Beemster Lake.

Many mills were used for different purposes at different times. A

*Windmills at work near Zaandijk, Netherlands.*

53 ❧

snuff mill called The Householder, for example, now grinds spice to make mustard. And a windmill called The Cat later became a prosperous merchant's home. Since his departure The Cat has become a museum for windmill memorabilia.

Before steam and electric power began to reduce Holland's need for the wind's power, the country had a windmill of some sort for every two thousand persons. A picture of Leiden drawn in 1647 shows its western wall lined with sails, and other towns were protected likewise. In the meadows between them, more sails revolved day and night as capriciously as the wind blew, and great clusters of whirling sails greeted seamen in busy ports.

Watching those windmills do so many things prompted ships' carpenters to hammer windmills together in their spare time. On the deck of a lumber lugger, a crude wind machine often pumped water out of the hold. In the 1970s when John Langfeldt was ninety, he could still describe one of those shipboard windmills vividly. When only fourteen, Langfeldt was a cabin boy on a Norwegian sailing ship hauling lumber to the Netherlands. On one voyage it ran into a storm so fierce that the older men swore the ship would have sunk if there had been no windmill on deck to help them remove water.

### PROUD PEOPLE

Life in western Europe became more exciting and strenuous in the century in which Leeghwater was born. New lands were discovered, new skills acquired, and new industries born. Artists often have told us things about the environment in which they lived that cameras have not recorded. Rembrandt was but one of many painters who portrayed windmills built by the Dutch of Leeghwater's time. At the Louvre, in Joachim Paintier's "Temptation of Saint Anthony," there's a windmill beside a river. In the Brussels Museum, a windmill dominates the background of Pieter Brueghel's "Wedding Procession."

The first settlers on our eastern seaboard erected copies of windmills like those that they might have seen when they were children. Today most people call every machine with four big rectangular sails a "Dutch" windmill. You no longer have to go to Europe to see a

*Like many Michigan citizens, De Zwann came to America from Holland.*

55 &

fine example of a real one in action. You need only go to Holland, Michigan, where an outside winder called De Zwann has been working for ten years.

De Zwann was built in North Brabant, where it first ground grain two hundred years ago. When the Dutch government granted special permission for the windmill to be brought to western Michigan in 1965, Jan Mendendorp, an experienced millwright, came along to supervise its reconstruction and operation. Nearly every large part except the millstones is wood. Although some of the original thirty-foot beams had to be replaced, most of the wooden machinery in the tower could still be used.

Descendants of Dutch settlers in Michigan had De Zwann reassembled on a mound of earth ten feet high in a municipal park that was once as swampy as the lowlands near the North Sea. The brick base is octagonal, two storys high, and was originally used for storage. The siding of the tower in which the milling is done contains thirty thousand cedar shingles. The cap is connected to a capstan wheel on the balcony by a triangular tailpiece, and the sails soar as high as a 12-story building over a park that is now called Windmill Island.

Everything possible is done the traditional way: the big wooden gears are lubricated with beeswax. Grain is lifted up to a floor near the cap, and from there it flows down to a meal floor where the miller manages the whole business. From the balcony he can pull a rope to apply a brake, stop the mill, and tie the sails down in foul weather.

The sails are tilted heavenward, and each one is angled slightly to intercept the wind obliquely. The sails turn when the wind blows between 11 and 15 m.p.h., revolving about 20 times a minute to turn the 11-foot gear and brake wheel in the cap. De Zwann is a well-known tourist attraction throughout its area. It has drawn millions of visitors to the park in Holland, Michigan.

Dutchmen remember the time when this mill was built as their country's golden age. Their explorers found no gold or silver mines as rich as those that the Spanish found in the New World. But Dutch sailing ships carried half of the sea trade of Europe. Although smaller than Spain, Holland became a greater industrial center with more prestige among nations.

# *The English Windmills*

By God's fair air,
   I grind ye grain,
Make good prayer,
   When bread ye gain.
LINES IN AN ENGLISH MILL

FOR SEVERAL CENTURIES windmills were as plentiful and often as useful in the lowlands of the British Isles as in The Netherlands. Their number was once estimated as high as ten thousand, and Essex was said to have as many windmills as square miles.

In *Henry IV* at the Globe Theater, Shallow reminded Falstaff of "a merry night" that they had lain in a windmill in Saint George's Field. "No more of that, Good Master Shallow," Falstaff replied, "no more of that." But when the show was on the road later, the actors learned to omit those lines; most windmills in the fortress built by nature were places for men to work in, not cavort, yet they were nearly always welcome sights.

Before 1830 a prolific writer, William Cobbett, told of seeing seventeen windmills while standing on one place. They were all white, their sails were black, and he thought "their twirling together added greatly to the beauty of the scene, which, having the broad and beautiful arm of the sea on one hand, and fields and meadows, studded with farm houses, on the other, appeared to be the most beautiful sight of the kind that [he] had ever beheld."

Robert Louis Stevenson could think of "few merrier spectacles than that of many windmills bickering together in a fresh breeze in a woody country; their halting alacrity of movement, their pleasant business, making bread all day long with uncouth gesticulations, their air, gigantically human, as of a creature half alive." Another commentator found windmills musical. When their stout arms bestirred themselves, he reported, "quasi-human whimpers would rise in weary protest, but once a steady rhythm was achieved, the mill in brighter mood, would chatter with a lively treble against the rumbling bass of the revolving stones."

Andrew Myllar, Scotland's first printer, had a windmill in his colophon in 1506. Merchandise of all kinds could be decorated with pictures of tall towers and long sails. Fine English bibles and psalters contained miniature views of post mills. A windmill even adorned a scene depicting the discovery of the Infant Moses, and another appeared in an illustration for an early edition of Boccaccio's *Decameron.*

Both on the moors and in the growing towns, the windmills toiled while stout Englishmen hoisted "windmill drinking cups" in their taverns. Those cups, used in the 1600s, had rounded bottoms and would tip over if set back on a bar before they were drained. After emptying one a man could show off by blowing into a tube to spin a tiny windmill on one side of his cup.

The British built bigger and stronger windmills as they gained experience. Their millwrights were among the first builders to substitute metal for wooden parts and to use roller and ball bearings. With crude and simple tools they solved many construction problems. Sometimes they drew plans for large mills on shop floors, so that inexperienced workmen could cut out pieces by following lines on the floor, like actors rehearsing for a play.

The largest tower mill in England was erected at Yarmouth in 1810; it was 122 feet high, its round base was 40 feet wide, and the walls were up to 3 feet thick. By then millwrights had learned to combine the skills of carpenters, masons, mechanics, blacksmiths, and architects, and to approach problems the way engineers have since been taught.

In one of the old fairy tales the Grimm brothers collected, a jack-of-all-trades found a man who could spin the sails of seven windmills two miles away by blowing one nostril. English millwrights disdained such fantasies. They concentrated on making sails easier for the wind to whirl and for ordinary men to manage. Wherever the windmills appeared, they hastened an industrial revolution.

John Constable's father owned a windmill, and the artist kept a wooden model of one in the studio where he painted his landscapes. His and other artists' work encouraged patriotic Englishmen to treasure and preserve windmills. At North Leverton in Nottingham, not far from Manchester and Birmingham, a mill with four big sails continued to produce flour in the 1970s alongside larger mills belching black smoke into the sky.

Some men still yearn to live in a windmill, but a beautiful tower may be as deceptive as some modern architecture. It's hard to fit square furniture into a round room and hang pictures on a wall that leans inward. The maid in one windmill home could only get in and out of her bedroom through a trapdoor in the master's room. Hilaire Belloc, a nostalgic man of letters, lived for nearly fifty years in a windmill where he could grind grain grown nearby into flour for his bread. But he soon tired of that flour, and often griped about the cost of repairing his home after storms.

Despite these disadvantages, though, windmills helped to instill the love for England that Benjamin Gough felt when he wrote:

> Ye happy homesteads, and broad orchards hail!
> The cheerful windmill, and the fields of corn,
> And fragrant hop with aromatic scent!
> Here would I live, and die where I was born,
> On some sequestered hill in lovely Kent.

### EXPERIMENTAL ENGINEERING

Most of the early British millwrights proudly shared their know-how with others, but some were extremely secretive. When Boaz Medhurst, one of the most careful builders, was about to mortice long masts to support sails, he locked everyone else out of his shop. Curious neighbors who peeked through a window at him reported that the millwright never looked so wise as when he was precisely setting out those mortices.

John Smeaton in Yorkshire, who claimed to be the first builder to use cast iron in millwork, invented an iron cross to attach masts to a driving shaft. This made it easier to put long sails in a rotor. It became known as a Lincolnshire cross when used on many mills there and in British colonies overseas. Smeaton published an *Experimental Enquiry* about his innovations and observations in the 1700s in which he laid down rules for designing windmill sails that engineers still go by. He pointed out, for instance, that:

1. The velocity of the tip of a sail can be almost directly proportional to the speed of the wind.

2. The maximum load on the sail can be nearly proportional to the wind's velocity squared.

3. The maximum power developed can be nearly proportional to the cube of the wind's velocity.

Edmund Lee, a blacksmith, found in 1745 that he could keep the big sails of a windmill facing the wind with a tiny "fantail." This was a secondary rotor with a few blades that he attached to the tower at a right angle to the main sails. When the main sails were squarely in the wind's eye the fantail was not, and it turned only slowly. But when the wind veered, the fantail began to whirl faster. It then could tug the big wheel to the right or left to keep the wind whirling it.

Lee's fantail acted similarly to a feedback device—the first of a series of inventions more portentous than anyone then suspected. Like a thermostat in a modern home, a fantail sensed a change in the atmosphere and adjusted a machine accordingly. A famous teacher, Professor Norbert Wiener defined feedback devices in the 1940s as means of "controlling a system by reinserting into it the results of a past performance." A feedback device enabled James Watt to make his steam engine run smoothly, and he has often been credited with having the basic idea first, but he got it from his country's windmills.

Millwrights put fantails on many windmills because they could act quicker than a man could hitch a horse to a tailpole to turn big sails in a different direction. Fantails were not the only way in which British builders strove to make windmills behave as though they had a bit of horse sense. Two other inventors tried to improve windmills with systems of metal springs while Americans were busy winning their freedom from England.

In 1772 Andrew Meikle, one of Watt's fellow Scots, substituted wooden slats like those of a venetian blind for sheets of canvas on the arms of a windmill. Those slats were connected to a bar alongside the mast and held in place by a spring. When the tension on the spring was right, it would let the slats open when the wind gained speed, to keep it from whirling the sails too fast, and it would pull the slats closed again when the wind slackened. Meikle's spring sail, however, yielded less power than a good common sail; the mechanism on each arm had to be adjusted separately by hand with an iron crank, and it often misbehaved in wet weather.

In 1789 Captain Stephen Hooper (whose horizontal mill made out of a packing box was mentioned earlier) patented another idea that involved a spring. He substituted a blind on rollers for wooden slats in a windmill sail. Hooper's "roller reefing sail" was often troublesome, too, and the captain acknowledged his problems with this invention when he advertised it in *The Kentish Gazette* by stating: "As many persons may prefer having the vanes made by their own millwrights, the Patentee will grant liberty for same on receiving a premium of ten guineas for sails not exceeding 30 feet in length, and four and six feet in width; and in proportion for those on an extended or contracted scale."

Eighteen years later Sir William Cubitt combined Meikle's and Hooper's ideas in a "patent sail" that was more successful. He had worked in his father's windmill, was apprenticed to a cabinetmaker, and became one of the most relentless students of sails who ever lived. Beside his deathbed, he had a tiny model windmill that he could make spin by pulling a string.

The vanes on Sir William's "patent sail" were controlled by bars attached to a "spider" in front of the shaft to which the sails were attached. His windshaft was hollow and a rod ran through it to a centrifugal device. In it a weight rose or fell when the wind whirled the sails faster or slower, and thus kept the vanes adjusted so that the windmill would run more steadily than the wind blew.

Thomas Mead had patented another mechanism in 1787 that was possibly more eye-catching. It had two weights suspended alongside an upright shaft. When the shaft revolved, centrifugal force threw the weights out farther as the shaft gained speed; they fell back closer to the shaft when it turned slowly again. Mead's invention was called a "lift tenter," because it was used in a windmill to regulate the rate at which grain flowed between millstones.

Sir William Cubitt's sail appeared on windmills in Denmark and

Germany as well as the British Isles, but never became popular in the Netherlands. It was a governor like Mead's that became famous when it appeared on Watt's steam engine. Watt considered mill-wrights "a contentious lot," and gave them scant credit for developing a technique that he could use. One of his associates, Andrew Moulton, saw Mead's "lift tenter" in a windmill where a steam engine was being installed. The builders were having trouble making the new engine run smoothly. So Moulton went back to the shop and Watt's assistants produced a governor similar to the "lift tenter" for steam engines.

Weights on a whirling upright spindle rose and fell from then on to control valves on steam lines. The governor was the most brightly polished mechanism on many steam engines, delightful to watch and easy to understand. More people saw it than studied the earlier governors on windmills, and Watt's version of Mead's invention was widely acclaimed as the first feedback device—but Watt's company never patented it.

In a scholarly work only a few years ago entitled *The Origins of Feedback Control*, Otto Mayr reminded engineers of the millwrights' contributions two centuries ago to technological progress. The feedback idea now has been widely used in electrical as well as mechanical inventions. An electronic feedback device steadies the picture you see on your television set, and related feedback systems have made our worldwide high-speed communication systems possible.

Feedback controls are not the only modern device that windmills anticipated. The flaps on the wings of our airplanes are quite similar to an "air brake" put on a windmill sail long before flying became safe. It was invented by an Englishman named Robert Catchpole. While one of sixteen workmen employed to build a big English mill in the 1860s, Catchpole put a long shutter parallel to the pole that supported a sail. That shutter could be closed or opened to increase or decrease the width of the sail, and it was called—more appropriately than people may have realized then—a "skyscraper."

## THE MILLWRIGHTS' MASTERPIECES

After studying many historic windmills, R. J. de Little recently wrote that "the absolute pinnacle of design was probably

reached in England" in the eighteenth century, when millers began
to use steam engines and rollers. At first the steam engines were often
only an auxiliary source of power for the big windmills that everyone
admired. Squire Roffey wrote in a parish magazine:

> The Mill with its sails like a sentinel stood
> On the crest of the hill o'erlooking the mere:
> And its timbers resisted, like British oak should,
> The storms and the tempests of many a year.

At Stalham in Norfolk, Samuel Cooke had the first windmill fitted
with Sir William Cubitt's patent sails. In the early 1800s, bands of
jobless workers called Luddites tried to halt the industrial revolution
by wrecking the new factories and machinery. One night in those
days, when Cooke climbed to the top of his mill, he saw thirteen fires
set by the angry workmen. Cooke's mill, however, was never threat-
ened; a steam engine was put in it in 1848, and it stood until 1903
when fire destroyed it.

At Canterbury the Holman brothers went on building windmills
so splendid that they became world famous in the latter half of the
nineteenth century, despite the increasing use of Watt's coal-burning
engines to grind grain. Sir Moses Montefiore commissioned the Holmans to build a windmill at the foot of Mount Zion in the 1850s.
He had found the poverty he saw in the Holy Land appalling, and
this seemed a sensible venture in foreign aid.

The Holmans made the parts in Canterbury, shipped them to
Palestine, and hired Arabs to help assemble the machine. But the
Arabs so relished the taste of oil that they insisted on licking it off the
bearings. Noting their aversion to pork, the engineers then hung a
pig's leg in a barrel of oil and dunked the bearings in it. That stopped
the licking, and the windmill served the people of Jerusalem for several years before it was converted into a studio and home.

Industrialization forced England to import more grain, and increased unemployment. Lord Darling wondered in *The London
Sunday Times*:

> Is it more cheap, upon the whole,
>     To buy abroad our grain,
> Let Labour languish on the dole,
>     Or windmills work again?

But the many improvements in windmills had made building one so expensive that only rich men could order one constructed.

The new steam-powered mills with rollers produced whiter flour than came from between millstones, and most people preferred it. Joseph Rank, one of England's leading millers, began with a small windmill. But when he saw a steel rolling mill for the first time in 1883, he acquired one like it that could produce more sacks of flour an hour than his windmill. Other millers sometimes found it profitable to use the wind's power only to grind feed for cattle. But some British farmers disliked the new steam mills' products so much that they pooled their savings to erect windmills to grind their grain the way they liked it.

Thieves broke into one of those windmills so often that the farmers began taking turns standing guard in it at night, and to get a bit of rest they devised a novel trap for an intruder. They stacked dusty, empty sacks over the entry in such a way that pulling a rope would let them tumble down to enmesh a thief. This trap was sprung only once, and the man caught turned out to be the most notorious smuggler in England. Suspecting that he had been hiding treasures of some kind in the windmill, the farmers then searched every cranny of it, but found nothing worth their trouble.

Over the years, windmills nearly disappeared from the British Isles. Many were lost in the First World War. The Holman brothers' business was continued until 1929. The last windmill that the firm produced was an aerogenerator. British experimenters in this century have since produced several novel devices to use both solar heat and windpower. A machine rated at 100 kilowatts that was built in the 1950s was especially interesting; air was sucked into a handsome tower to run a turbine, and hurled out of the tips of hollow blades by centrifugal force.

British engineers are still noted for their knowledge of historic windmills and efforts to preserve them. In 1960 R. J. de Little found really old windmills working in only about eight places in England, but there are more today. Some are now run only to entertain their owners and visitors, but a few were operated again as businesses in the 1970s. This has happened on a small scale in the United States, too, and windmill watchers in both countries can again enjoy the sight of big sails revolving in bright sunlight on fair horizons.

## DORIS DUKE'S WINDMILL

Near Newport, Rhode Island, Miss Doris Duke has had a fine example of a British type of mill reconstructed to be run again the way it did when it was built in the early 1800s. Miss Duke is an elderly tobacco heiress who still lives in one of Newport's palatial homes. She established a foundation in the 1960s, with herself as president and Mrs. Jacqueline Onassis as vice president, to make Newport a unique American community. They envisioned a city as charming as the historic section of Williamsburg, Virginia, and as vibrant as Georgetown in Washington, D.C.

For the remaking of Newport, Francis Adams Comstock, a former director of the Princeton School of Architecture, came out of retirement to direct the Newport Restoration Foundation's work. The foundation has since restored and rented to appreciative tenants many of the small houses built in downtown Newport in the eighteenth century, and one of Comstock's first and most challenging projects was to place an English smock mill in an idyllic setting a few miles north of the old town.

The windmill he chose has been moved several times. From Warren, Rhode Island, where it first ground corn, it was hauled across the ice in Mount Hope Bay one winter to Fall River, and from there to other sites later. Despite that wear and tear the windmill still contains many of the original oak timbers and looks as if it had been built only yesterday.

The sails of the old mill at Newport draw attention to a cluster of little red frame buildings on the Prescott Farm, near the house where a general commanding British troops surrendered to about forty American rebels on July 10, 1772. The dome on top of the windmill's octagonal wooden tower is 58 feet high. Four rectangular sails soar 72 feet into the air after sweeping within reach of a man on the ground. The tower has four floors. A pine pole a foot and a half thick whirls to transmit energy from the wind to big gears on the ground floor.

Those gears send that energy back to millstones on the second floor. Each stone in one pair weighs a ton and each stone in a second weighs a ton and a half. Moving this heavy old mill to its present site was a stupendous job that took nearly two years. No available helicopter could lift the whole thing. If set on a truck to be hauled down the highway, power lines would have to be cut. A Raytheon plant nearby could only tolerate an interruption in its power supply on a Sunday, and a state law forbade the use of the road to move a building then. So the tower had to be cut in half, placed on two trucks, and put together again on the Prescott farm.

Another law required that wooden funnels be lined with stainless steel for sanitary reasons before any corn meal ground by the old stones could be sold. Metal gears also had to be substituted for old wooden gears that had lost too many teeth to be useful; no one could be found who could make new wooden teeth which would mesh properly. Solutions to these problems made Miss Duke's machine a unique work of art.

When the windmill was finally operable, the restoration foundation advertised in vain both in newspapers and on television for a miller able and willing to run it. Fortunately for the project, a dairy farmer named George Kimball was in the neighborhood, came by, and stopped to see the windmill. Still in his twenties, Kimball was from Pittsfield, New Hampshire, where he had won national honors in 4-H Club contests and become a specialist in crop breeding. Comstock found Kimball so engaging and enthusiastic a chap that he persuaded him to learn how to run a windmill.

Kimball spent a month studying every detail of this mill and more time touring New England to question operators of gristmills still being used. They were all run by water power, and he soon discovered that using the wind was more precarious. The four 180-square-foot sails of Newport's reconstructed windmill had to be rolled tightly against the masts when he furled them. This meant that he had to climb part way up the lattice work on each mast. He fell while doing this, injured his back, and the formal opening of the Prescott Farm mill had to be postponed.

Nevertheless, Kimball moved with his wife and children into the restored miller's cottage a few yards from those sails to manage them. The wind determines his working hours. When it blows fifteen miles or more an hour, the sails revolve about twenty times a minute and the millstones spin five times as fast. But the wind loafs so long some summers that Kimball has often run out of corn meal to sell to visitors. To solve this problem Kimball persuaded Miss Duke to buy him a walk-in electric freezer that he can fill with meal ground in the spring when the wind blows more often.

*This rebuilt smock mill is a few miles north of Newport, Rhode Island.*

Kimball is also breeding White Cap Flint Corn back to seeds like those that the Indians gave the English colonists. This takes time, but he hopes to do likewise with rye and wheat. During the tourist season, he needs two helpers to answer visitors' questions, care for the grain he is growing, tend an herb garden, and keep up with his other chores. One of his young apprentices, Matthew Weaver, has also learned to run the windmill, and both men enjoy their jobs.

They admire the workmanship of the mill's builders and proudly point out how easily the distance between the millstones can be varied by a hundredth of an inch. They prepare for windy days by hoisting grain up to bins above the millstones from which they can let it flow down at whatever rate they choose. On one occasion they ground a thousand pounds of corn meal in two hours. When the mill is going full tilt the big metal gears sound like a locomotive racing by, and the whole tower vibrates. So much happens so fast that the miller and his apprentice listen as well as watch for something to slip, and keep everyone else out of the tower lest someone be injured.

Up the road a short way from Prescott Farm, on E. P. Boyd's place, there is another English smock mill of about the same size. Dense foliage hides it from summer tourists, but it looks very much the way the reconstructed mill did when Miss Duke bought it. Boyd's mill has been idle for many years, and the wind has ripped off its sails and torn shingles loose from the tower.

# PART TWO

# The Pioneers' Partners

# VII

# *America's First Windmills*

There's something in the wind.
THE COMEDY OF ERRORS, ACT III, SCENE 2

AS INEVITABLY AS gypsies, European post and
smock mills followed colonists to the New World. After the wind
blew Columbus across the ocean blue in 1492, settlements spread
northward from Central America to our country's great bays, river
mouths, and far inland.

In 1542 a Spaniard named Don Francisco Montayo chose the site
of a Mayan city called T'ho on the Yucatan peninsula for a town.
The settlers named it Mérida, which is pronounced in their language
as if it were *"merry*-da" in ours, and the wind has especially blessed
it. Rainwater seeps through the limestone plain on which Mérida
stands and replenishes underground pools and rivers. It was a good
place for cactus-like plants from which the Spanish made hemp, and
windmills pumped their water for them.

Mérida later became a provincial capital, then a city of 200,000,
and is now a base for tours of the ruins of Mayan civilization. Windmills have continued to deliver water to the citizens' homes in this
century. The water is crystal clear, and the air is so clean that visitors
from the United States have remembered Mérida as "the white city."

In Yucatan and throughout the New World today, of course,
factory-built machines outnumber the kind that Don Quixote as-
sailed. Although the Spanish built windmills in many places, they

71 ह

left few traces of them in our country. Explanations vary: The Spanish looked mainly for gold, silver, and souls to save. Wood was scarce at many mission sites; after going up the Rio Grande and exploring the plains where the buffalo roamed, Coronado reported "there is not any kind of wood in all those plains away from the gullies and rivers which are few." A Stanford Research Institute engineer who has searched many California missions for evidence of windmills in vain suspects the available wood was needed more urgently for other purposes. The Spanish held those missions only briefly, and may well have considered saving the Indians' souls more urgent than saving human energy.

The Danes in the Virgin Islands that the United States now possesses brought both slaves and windmills to the New World for their plantations. Slaves harvested the sugarcane and windmills ground it. The windmills never rebelled but the slaves did when crops failed in the 1700s. Before French troops arrived to help the Danish militia restore order, hungry slaves hacked twenty-five families to death.

Professor Terry G. Jordan, chairman of the North Texas State University's geography department, has mapped the sites of many early windmills in North America. Dots on that map show that Frenchmen began to build windmills on the banks of the St. Lawrence River in 1629, and built others between Lake Erie and Lake Huron. A French windmill also went up near St. Louis, and another on the Mississippi River's delta in the 1700s. A century-old sketch shows a battered windmill tower at Detroit that could still support sails in 1872.

Near Lake Winnipeg, in Canada's Red River Valley, colonists were unable to assemble parts of a windmill sent to them until a Scottish millwright arrived in 1845.

Less than a decade after Governor Yardley built an English mill in Virginia, another rose near Massachusetts Bay, and the Dutch built windmills at the mouth of the Hudson River. Swedes, Germans, Portuguese, and other nationalities increased the variety of early American windmills. On the Potomac River's bank, "Bluff's windmill" ground grain for many years about where the Kennedy Center for the Performing Arts stands in the District of Columbia today. Swedish settlers put up an especially famous multi-purpose wind engine at Lawrence, Kansas, in the 1800s.

✑§ 72

In 1715 "his Excellency the Palatine and the rest of the Lords Proprietors of Carolina" acted to encourage the construction of both water and windmills. For any person wishing to build a water mill the Surveyor General was authorized to lay out two acres of land, and for anyone who wanted a windmill half an acre, "in such manner as shall for that use be most convenient"—provided that it be paid for at a price set by four honest men of the neighborhood.

The waterfront street at Beaufort, North Carolina, had three windmills by 1839, one of which resembled a little corncrib mounted on a single post two feet in diameter. The writer who described it noted that masts salvaged from ships made good tailpoles for post mills. Horses hitched to them turned the sails in the desired direction on "mill rounds." Some rounds were also marked off with big rocks so that a miller without a horse could use a block and tackle.

Offshore islands had no rivers suitable for water mills, and windmills often appeared first there. A student of North Carolina's historical geography, Gary S. Dunbar, has located the sites of fourteen early windmills on the long sandy strips between False Cape and Cape Fear. The first one was probably built in 1700, and at least one old-timer was still running two hundred years later. Two windmills once stood on Cape Hatteras, the graveyard of many ships, and two more at Kitty Hawk, aviation's birthplace.

North of Chesapeake Bay water mills were more plentiful than windmills. One mill built in Camden County, New Jersey, in 1762 could be run either by wind or water, and the wind was seldom needed. There must have been many other wind and water mills in the area in those days, but when Harry B. Weiss, a writer for the New Jersey Agricultural Society, searched that state for early grain mills in the 1960s, he found none left that was driven by the wind.

New Jersey's finest windmill was probably the Great Western Flouring Mill built by a baker named James Edge. He hired two British millwrights in 1815 to construct it so that he could have flour for his bakery like his father had used in England. When a storm wrecked his mill's canvas sails they were replaced with sheets of metal. That mill stood across the Hudson from New York City where Jersey City is today until it was moved to Long Island, where it survived until 1970. Now Milford, New Jersey, has an "educational copy" of a beautiful old European windmill.

## ALL AROUND THE TOWN

New Amsterdam's first governor is credited with putting up three windmills. The Dutch went on to line Broadway with them from the Battery to Park Row. From a ship entering the big harbor then, the windmills sawing lumber and grinding grain made the new city look like a port in The Netherlands. The road to Peter Jansen Mesier's mill at the foot of Cortlandt Street was called Old Mill Lane.

Ship captains watched the windmills on the waterfront to see how to set their sails when entering or leaving the harbor. Ferries ran from Manhattan to Long Island "daily except when windmills on the opposite shore have taken down their sails." At such times the law forbade anyone to cross the East River, even in a rowboat.

Fine flour was produced by a process called bolting, in which grist was passed through sieves after it came from between millstones. In 1665 when Sir Edmund Andros was governor of New York, bolting was forbidden outside of the city. That short-lived attempt to maintain a monopoly accounts for the four big windmill sails and two barrels of flour still in New York City's official seal.

Washington Irving reported in *Knickerbocker's History of New York* (1809) that "certain wise old burghers of the Manhattoes, skillful in expounding signs and mysteries, consider the early intrusion of the wind-mill into the escutcheon of the city, which before had been occupied by a beaver, as portentous of its fortune, when the quiet Dutchmen would be elbowed aside by the enterprising Yankee, and patient industry overtopped by windy speculation."

Indians came from afar to see the Dutch windmills, and sometimes loitered in the city for days to study them. Although the millwrights no doubt believed New York would be a great place to live if its people ever finished building it, the Indians were dubious. They set fire to some of the first windmills that the Dutch built farther up the Hudson Valley. Since then Indians have helped to fill Manhattan with tall buildings rather than windmill towers.

The Dutch windmills' brief tenure in New York did not retard their progress westward. In Illinois one was built in a cemetery. At Burlington in Iowa a windmill's sails beckoned people on across the Mississippi River, and the tall towers of two powerful Dutch windmills still stand in San Francisco's Golden Gate Park (awaiting restoration of their sails).

Dutch millwrights, however, were not allowed to practice their trade east of Brooklyn and Queens. The English settlers on Long Island insisted on having English windmills. There the first windmills were usually smock mills about three stories high. The frameworks were generally oak, the floorboards pine and the shingles cedar. The caps were often shaped like boats.

Many of those windmills have been well preserved, although some are no longer on their original sites. The Long Island Historical Society has dated six of them as follows: *East Hampton*—Gardiner mill (1771), Pantigo mill (1801), Hook mill (1806); *Southampton* —Shinnecock mill (1697); *Bridgehampton*—Bridgehampton mill (1820); *Shelter Island*—Sylvester mill (1810).

Rex Wailes, the technical advisor of the British Society for the Preservation of Ancient Buildings, toured Long Island in 1962 to inspect, measure, and photograph most of the early windmills still upright. He reported that East Hampton had as many left as any village in England, and found its windmills especially interesting. The Kentish folk who founded East Hampton in 1640 had first called it Maidstone. Elm trees lined its wide half-mile-long main street between two village greens where windmills rose. For several generations the Dominy family built and ran windmills in East Hampton.

One of them, the Hook mill, was built from oak felled on Gardiner's Island, a breeding place for fish hawks where Captain Kidd's pirates hid their loot. John Howard Payne wrote "Home Sweet Home" about a cottage in East Hampton, and New Yorkers began to flock there early in this century to escape from their city's hot and dirty streets in summer. Artists among them so admired the three old windmills still standing that the village acquired possession of two of them.

The town fathers had new teeth put in the Hook mill's gears, gave it a new dress, and made "Puff" Dominy its custodian. He was a great grandson of one of the island's famous millwrights and had learned to run that mill when he was a boy. "Puff" enjoyed showing visitors how it was constructed and explaining that "like a sailboat we always sail close-hauled to wind'ard, never to leeward."

*Windmills around East Hampton, Long Island, about 1820.*

In coastal communities, early American windmills were used to signal ships and sharpen whaling harpoons, as well as to run saws and turn millstones. One, built on Shelter Island in 1810 could still be used to grind grain during a world war more than a century later. Old families and public-spirited groups have saved several Long Island windmills that have colorful histories. Others have been built mainly for show, and some have been moved so often that it is a wonder anything is left of them.

Prof. G. W. Pierson of New Haven intervened to save a famous windmill on Montauk Point, Long Island's eastern tip. When he read that it was for sale, he set out to raise a fund from friends to buy and move it to nearby Wainscott. Before he had collected enough money, the U.S. Army fenced off Montauk Point for military use, and decided to tear the windmill down. The professor then got it for nothing by promising to move it quickly. But he hired a contractor who wrecked the mill when he lifted it off its base. The professor and friends had to pay out $2,930.30 to get the pieces hauled to Wainscott and put together again.

Few big machines have been as coveted as windmills. Brooklyn's Prospect Park bought a windmill in Bridgehampton some years ago to move to the city. But that turned out to be too costly, so it was left at Bridgehampton, and later put to work again there.

### RHODE ISLAND'S TREASURES

Half a century ago *The New York Times* called the land around Long Island Sound "windmill country." Few windmills have survived real estate developments on the Connecticut shore, but several fine old English mills have been restored and reconstructed in Rhode Island and on Cape Cod. The family that built the first European residence at Newport also erected a windmill in 1639. Others soon adorned nearly every hilltop in the vicinity, and the most puzzling old tower in America still stands in a tiny Newport park on Mill Street.

The stone structure is a plain wide cylinder supported by eight pillars. The arches between the pillars are high enough for a person to enter from any side of it. The pillars are on true compass points, and the facing stones could not have been found anywhere near Newport when it was built. The 26-foot-high tower is similar to very old Norse churches, and there are indications in it that it once contained a fireplace and altar. Several generations of researchers consequently have contended that the tower was a place of worship built by the Vikings in the eleventh or twelfth centuries. This possibility has stimulated the imaginations of writers including James Fenimore Cooper and Henry Wadsworth Longfellow. Schoolbooks still contain romantic legends about it. In 1785 the relic was used as a lookout tower, and in the 1800s the U.S. Naval Academy was nearby; but little is certain about the tower's early history.

The evidence that English colonists built the odd tower to support sails for a windmill in the seventeenth century is as convincing as any that the Vikings left it there centuries earlier. In 1677 Governor Benedict Arnold (whose descendant later surrendered West Point to the British) mentioned a stone-built windmill in his will. The colonial governor is known to have had a hand in the construction of a wooden windmill on a nearby island at Jamestown. Newport's mysterious tower closely resembles that of a very old windmill still standing in England that the governor could have seen when he was a boy. In 1714, moreover, the oldest man in Newport asserted that this tower had once held up sails so big that it took a team of oxen to redirect them when the wind changed.

Artists and writers now share the Newport breezes with the heirs and heiresses of industrial tycoons and the owners and crews of beautiful yachts competing for the America's Cup. By the 1970s many of the people who live there were inclined to believe that the tower on Mill Street was built for a windmill. But then a medical geneticist whose hobby was studying ancient Norse churches showed up. He examined it and was so convinced the Vikings built it that he proposed that a medieval church be built around it to celebrate the nation's bicentennial. This revived the local historians' debates at cocktail parties about the strange tower's origins.

More is known about a windmill that Governor Arnold and other wealthy colonists paid to have built at Jamestown, near Newport. They had bought the narrow island in Conanicut Bay in 1663 from an Indian chief who lived there in the summer. They paid him a hundred pounds plus some incidental gifts for that narrow sliver of land, and promptly ordered a windmill built to serve the settlers. It was replaced in 1787 by another one, which was still erect when Jamestown, like Newport, became a summer resort for affluent Americans. In the 1900s this structure became the object of the best-recorded early effort to save an old American windmill.

The mill, on a hillside near the road into the village, had not been run for nearly ten years. A family living on the place had let it deteriorate when Miss Jane Eliza Woodin wrote:

> My mind reverts with pleasure for a while
> To that old Wind Mill on our Sea Girt isle
> Which erst has ground from corn and golden grain
> A healthy nutriment for brawn and brain.

Mrs. Frank S. Rosengarten of Philadelphia, and other ladies summering in Jamestown who shared Miss Woodin's sentiments, found that they could buy the mill for $300 and get it restored for $400. A whist party was given to start raising the money. Other social affairs followed, and in 1912 the restored mill was entrusted to the Jamestown Historical Society.

For the last twenty years Mr. and Mrs. Robert McCallen have lived in the miller's house by the side of the road to protect the windmill from vandals. Mrs. McCallen is a tiny lady with a winsome smile who is familiar with every detail of the mill's history. Both the front and back steps to her charming home were once millstones, like those used for tombstones on millers' graves. A sudden powerful twister in the spring of 1974 wrecked all four sails and the windmill's arms had to be removed, but by then the historical society had no trouble finding funds to replace them.

The Jamestown mill resembles Miss Doris Duke's machine described in Chapter VI, but it is much smaller and contains more of the original parts. The octagonal wooden tower is less than twenty feet wide and has never been moved. Tiny windows spiral up the sides to a dome-shaped bonnet. Bonnet and sails were originally turned by oxen, but can now be turned by means of a rope and wheel behind the sails.

The mill's single set of 5½-foot-wide stones is on the ground floor. The runner revolves eleven times as fast as the sails when 540 square feet of canvas are spread on them. Even so, the records show, it usually took more than an hour to grind a bushel of corn between those millstones.

Mrs. McCallen ascribes much of this windmill's popularity to "Johnny Cake Days." Members of the Society for the Propagation of the Johnny Cake Tradition in Rhode Island contend that "the best article of farinacious food ever tasted by mortal men" was baked from white corn meal ground between granite stones. The cakes used to be baked on a red oak barrel head supported by a flatiron in front of the glowing coals of a hardwood fire. After Jamestown's mill was restored, Rhode Island women set up stoves one day every summer in a field beside it to cook Johnny Cakes with meal warmed by its millstones. Some folks say such cakes "make city bread taste like cardboard," and Rhode Islanders have served them to guests from all parts of the country.

Steam and electrical power replaced windmills for grinding grain in the United States earlier than they did in Europe, and more completely. Most wooden contrivances were also discarded sooner here. Hence, to J. Kenneth Major, a British architect and authority on milling, some of our oldest restored mills now seem even "more primitive" than those preserved in other countries.

## CAPE COD'S BIG TURTLES

The first windmill in Massachusetts ground corn for housewives between Watertown and Harvard College at Cambridge. There the mill responded so poorly to the westerly wind that it was moved in 1632 to the Boston waterfront. Additional windmills soon bobbed up in Charlestown, Scituate, Salem, Portsmouth, and other towns. Toolmakers used them to produce cranberry rakes for farmers and shovels for the first railroad builders. At West Dedham, Edward Glover and his son produced tools with energy from the sails on the roof of their shop until 1905, when the firm acquired a gasoline engine.

On Cape Cod the abundant windpower helped to meet two of the settlers' basic requirements, bread and salt. So little fresh water flowed into the ocean from that long hooked cape that it was a fine place to obtain marine salt. Extracting it from sea water became a major industry when salt was scarce in 1776, and the salt business boomed for the next half century. Captain John Sears started it by using the wind to pump salty water into big wooden vats, where the sun could evaporate it and leave chunks of salt.

The cod and mackerel fishermen needed lots of salt to preserve their catch. The cape's solar salt works were most productive in the summer when the men were busy fishing, but women and children had time to run the plants then. To keep out dew and rain big wooden lids were swung over the evaporating vats, and the lids made the vats look like huge turtles to Henry Thoreau.

By the 1830s, 442 salt works were producing half a million bushels a year. They lined the waterfront all the way to the tip of the cape where seventy-eight were clustered around Provincetown. Marine salt brought up to seven or eight dollars a bushel until salt deposits were discovered inland. The price fell then to about a dollar a bushel, and when Thoreau toured the cape a few years later he found the salt works "fast going into decay." They are all gone now.

Thoreau noticed the big weather-stained windmills grinding grain as well as the salt vats on the Cape Cod shore. His descriptions of such attractions in the 1800s helped later to lure more travelers to the cape's sand dunes and beaches—and thus to launch a new industry.

In 1909 a contributor to *American Machinist* magazine inspected Cape Cod's remaining windmills and was amazed, both by the builders' ingenuity and by the durability of the old English smock mills. He found tough pieces of oak still held together firmly by cotter pins rather than by nuts and bolts. Blacksmiths had made those pins by thrusting pieces of old horseshoes through holes that they forged in iron rods. The reporter also admired an eight-foot wooden gear with teeth made by driving square pegs into its round rim, and discovered that a woodpecker had drilled a hole in a solid oak driving shaft.

Several Cape Cod windmills like those that Thoreau saw have been well preserved. Roadside stands sell picture postcards of the old windmills, and the guides on sightseeing buses point them out. Sandwich, the oldest town on the cape, has the oldest windmill that you can count on seeing in operation during the tourist season. It can be run by a hidden electric motor whenever the wind is not strong enough to drive the sails. This "Old East Mill" is on the Heritage Plantation,

*The Old East Mill is now at the Heritage Plantation at Sandwich.*

a park museum for relics of simpler times. A non-profit corporation opened this showplace in 1969 as a memorial to Josiah Kirby Lilly, Jr., an ardent collector of old coins and other Americana.

A syndicate of five men had built the windmill in 1800 to grace the town of Orleans. Oak and pine, discarded when a new porch was put on the Congregational Meeting House, went into the mill. Deacon Abner Freeman bought the other owners' rights to the mill for $83.31 in 1811. He sold it to a minister a few years later when coins were scarce, and the minister kept two quarts of grain from every bushel it ground as a toll.

In Orleans the mill was used to grind chunks of salt in pots of sea water alongside the millstones. Eventually it was moved to Rock Harbor, where the Long Island–Connecticut–New York packet boats docked. A cobbler named Isaac Snow ran it there until he got a Civil War pension. A legendary sea captain, Joseph Taylor, then bought the mill on his retirement from the sea and used it to grind grain until 1893.

When moved to the Heritage Plantation at Sandwich, the Old East Mill was rebuilt to serve as an educational exhibit, and C. Malcolm Johnson, a retired spring-water dealer who has read widely about windmills, became the miller. He has become a popular lecturer on Cape Cod who will gladly tell you more about windmills than you ever thought you wanted to know.

Johnson likens the profile of the mill he runs to that of "a sea captain dressed for rainy weather." Its turret sits loosely in a groove, and was originally turned with a long tailpole that is no longer needed. The sails are strips of canvas spread on four arms attached by a Lincolnshire cross to an oak driving shaft. They must be furled by hand, but the miller now leaves them spread most of the time. He can stop the mill to take the sails in whenever he likes, yet keep his millstones turning by switching on its new electric motor.

The millstones are on the ground floor. You can step right in and look up through an opening in the ceiling to see big brightly polished wooden gears turning in the cap. A spindle gear and an iron rod called a damsel connect the upper millstone to the driving shaft. You can smell the fresh corn meal thrown out from between the stones while observing the whole operation. The machinery creaks monotonously when run by electrical power, and seems less musical than it does when the wind takes over.

Cape Cod has older English smock mills than the Old East Mill; one at Eastham was built in 1793, one at Brewster in 1795, and another at Chatham in 1797. A windmill erected in 1742 still stands on Nantucket Island. These older mills are never run by electric power like the one at Heritage Plantation, however. There, a visitor can see, smell, and hear everything whenever he happens to arrive. In a 20-m.p.h. wind, the miller will assure you, this one can grind a bushel of corn in 20 minutes.

That isn't enough to interest today's big food processors, but seeing flour produced this way is so alluring a treat for tourists that both Yarmouth and Brewster set out in the 1970s to refurbish their old windmills. At Yarmouth the town put up $30,000 to move and fully restore the Judah Baker mill on a popular beach. Vandals broke in and set fire to that mill before it was ready to be reopened, but firemen arrived in time to save much of it.

VIII

# The Windmill Factories

Of all the forces of nature, I should think the wind contains the greatest amount of power.

ABRAHAM LINCOLN IN 1859

IN THE UNITED STATES engineering was first taught only in military schools. We had no big scientific laboratories or institutes of technology in the first half of the nineteenth century, but the air was fresh and stimulating, and young men learned the industrial arts by watching older men. Mr. Lincoln's contemporaries put tires inflated with air around bicycle wheels, and before he became president he himself patented similar tubes to float riverboats off sandbars. Windmills helped to make possible increasingly complex and powerful machinery, which gradually transformed a meagerly developed rural country into a world power dominated by an urban industrial society.

On the western frontier an immigrant from Germany acquired an Iowa farm in 1855 on which he put up a 70-foot tower to support a wheel 17 feet wide for the wind to whirl. Machines like that were a challenge to other ambitious men.

Before the war between the states, a windmill with many more than the usual number of blades was patented and widely advertised,

# WIND MILLS.

## For PUMPING, &c.

### HALLADAY'S PATENT.

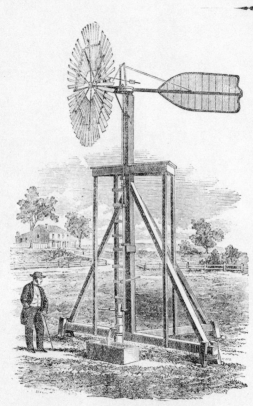

THE two essential points in a good Wind Mill are, that it should HEAD ITSELF QUICKLY TO THE WIND, and that its motion should be so "SELF REGULATED" as to prevent accident in storms, and preserve a nearly uniform speed of revolution in winds of different force.

The Mill shown fills these requirements and is substantially built.

The size and kind of PUMP will depend upon the size of mill, location, depth of well, &c.

The method of SUPPORTING THE MILL depends upon the location, and will rarely be the same in two places.

A CONSTANT SUPPLY OF WATER may be obtained for any ordinary purpose, by a proper regard to the size of tank, size of mill, and the force and continuance of wind.

The larger size mills, from 16 to 60 feet in diameter, are used for shelling corn, cutting fodder-grinding grain, and other farm operations.

Mills and Pumps can be seen at my Store, and information will be given, suited to each particular case, with drawing of framing &c., if desired.

An inspection of arrangements for Water Supply, Baths, Basins, Sinks, Water Closets, Bidets, &c., would repay a visit to

## W. G. RHOADS,

### 1221 MARKET STREET, PHILADELPHIA.

#### MACHINES FOR SUPPLYING AND FIXTURES FOR USING

## WATER, STEAM AND GAS.

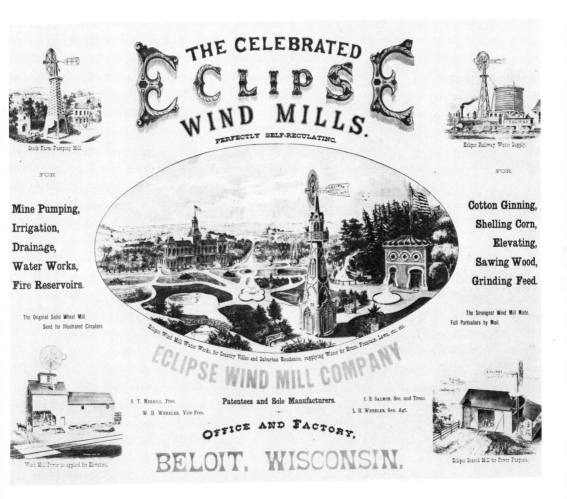

Advertisements for the two main rival windmill manufacturers of nineteenth-century America: Halladay (left) and Eclipse.

and in the next few decades Americans patented nearly a thousand more ideas for improving windmills. Scores of factories began to build them. The most successful manufacturers produced fans with many short solid blades for the wind to whirl vertically. New mechanisms kept the fans facing favorable breezes, slowed the wheels automatically when necessary, and protected them from the wind's occasional savagery.

In the Colonial period, most windmills were built to grind grain, but nearly all of the new windmills were designed to pump water. Their towers, which did not have to house millstones and other heavy machinery, were scrawnier and less attractive than those of the old Dutch and English post and smock mills, but the little fans drew as much power from the wind as was needed to fill a water tank. The small fans were less costly to produce and easier to erect than the old gristmills had been, and they required less supervision.

The hucksters for the factory-built machines soon found three big domestic markets for them. The railroads needed windmills to fill water tanks for the steam locomotives chugging across the continent. Well-to-do families everywhere bought them to have running water for bathrooms. And the cattlemen and homesteaders moving into the western plains bought them to irrigate arid land and to water livestock.

The *U.S. Statistical Abstract* of the Census of Manufactures in 1919 reported the windmill industry's growth in fifty years:

| DATE | EMPLOYEES | VALUE OF PRODUCTS |
|------|-----------|-------------------|
| 1879 | 596 | $1,011,000 |
| 1889 | 1110 | 2,475,000 |
| 1899 | 2045 | 4,354,000 |
| 1909 | 2337 | 6,677,000 |
| 1919 | 1932 | 9,933,000 |

In 1889 there were 77 windmill factories in the United States; by 1919 the number had dwindled to 31. The windmills that they sold produced hundreds of thousands of gallons of water every day without putting their owners to much trouble.

By 1903 windmills had become a significant factor in American exports. State Department consuls reported substantial demands for American-made windmills nearly everywhere except in Europe, and even there farmers had begun to prefer the new American machines. In Mexico, Edward Thompson took potential buyers to Mérida on the Yucatán peninsula to show them "two bright windmills of metal working merrily in a wild zephyr, while the ponderous wheels of the others were either sulking silently or turning croakingly."

Twenty-one million persons went to the 600-acre grounds of the World's Columbian Exposition on the shore of Lake Michigan in Chicago in 1893. There many people first saw Pullman sleeping cars, Edison's phonograph, and other new mechanical and electrical wonders. But the sunlight glinting from the sails of dozens of windmills pumping great streams of water from shallow wells around a lagoon made the wind machines one of the great fair's brightest sights. Those busy windmills not only pumped water, but also shucked corn, ran lathes and sewing machines, and showed how the wind could lighten other familiar chores.

"Each manufacturer claimed some superiority," a newspaper wrote of the machines exhibited in Chicago by American companies and the Dutch government. "Here a wheel would open to get more wind or shut against too much; one mill would go swiftly in the lightest breeze, another would work slowly in a hurricane."

Most of those machines had been made in small towns near Chicago—places like Batavia, Freeport, Waukegan and Sandwich in Illinois; Kendallville, Auburn, Butler, Elkhart and Mishawaka in Indiana; Beloit, Waupun, and Racine in Wisconsin; Springfield, and Napoleon in Ohio; and Kalamazoo in Michigan. Often the windmill factory was the first and for many years the only industry in those trading centers. It employed men whose help was no longer needed on nearby farms because of the increasing mechanization of agriculture, and their wages went into the tills of local merchants.

Some of the machines were so admired that they were pictured— like race horses, popular pugilists, and other celebrities of the day—on the shaving mugs that men chose from when they went to a barber shop. Workmen who had built those windmills followed them to Chicago to root for them the way people now do for local football teams. The competition between the windmills at the Exposition was as spirited as the simulated conflicts on the fair's midway between Buffalo Bill's cowboys and Indians. Civic pride in fact drove a few

fans to violence; they broke into the fairgrounds one night and tore down a strong contender for the prize to be awarded in a windmill contest the next day.

Few new industries had been born in so favorable an environment as the American windmill factories. But our country's history has been so poorly taught that few people now know of the windmills' feats at the Columbian Exposition. Not even a picture in the big Museum of Science and Industry on the site of that fair reminded visitors in 1973 of the windmills that once competed there with other inventions.

*Windmills competed at the World's Columbian Exposition, Chicago, 1893.*

### HALLADAY'S HINGES

The two inventors who were most widely acclaimed a century ago for giving birth to the American windmill factories lived far apart and had very different interests. They were Daniel Halladay, a young mechanic in Connecticut, and the Reverend Lawrence R. Wheeler, a missionary to the Indians in Wisconsin. Both men were famous only briefly. The genealogical files of the Connecticut Historical Society in Hartford contain no record of Halladay's existence. Reverend Wheeler became a prominent local businessman, but the company that popularized his invention bore other men's names.

Halladay had to be talked into building a windmill. John Burnham, a roving "pump doctor" in New England, had noticed what a nuisance having to furl and unfurl big sails for a wind-driven pump was to a farmer. Too many landlubbers' machines were being wrecked by the wind because their owners were too busy doing something else to pay attention to the wind's tricks. Burnham found Halladay running a little machine shop at Ellington, Connecticut, in the 1850s, and told him about the problem.

"I can invent a self-regulating windmill that will be safe from destruction in violent windstorms," Halladay is said to have declared, "but I don't know of a single man in the world who would want one." Burnham thought nearly everyone would want one if the new windmill did not cost too much, and persuaded Halladay to try.

The young mechanic had seen and studied the governors that kept the new steam engines from running wild by using centrifugal force to feed back a warning to a valve. James Watt's company had borrowed that idea from a mechanism used with millstones, and Halladay borrowed it back to regulate the rate at which the wind could whirl a vertical wheel around its hub.

The sails of the machine that he designed were thin wooden boards. Instead of attaching them to the hub, Halladay hinged them on a ring around the hub. He then connected them through a sliding collar on the shaft to a weight that rose or fell to vary the blades' angle, and therefore the surface that the wind struck directly when it gained or lost speed. When the wind was gentle, Halladay's fan wheel was nearly flat and the wind struck the broad side of every blade in it. But when the wind became fierce, the boards on the ring swung out of its way. The rotor then resembled a basket with an open bottom that a ferocious wind could blow through without harming.

Halladay's first fan wheel had only a few blades, but he increased the number in later models by hinging them on six or eight straight rods in a circle around the hub, allowing for a dozen or more blades. This so-called "sectional" wheel was nearly flat when it was "in sail"; it took the form of a cylinder when a strong wind caused the blades to tilt away from the wind putting the wheel "out of sail." Although it certainly was not as pretty as a rose, Halladay's fan wheel was dubbed a "rosette," because its blades seemed to respond to nature somewhat like the petals of a flower. To keep this wooden wheel in

*Halladay-type windmill "in sail" (left) and "out of sail."*

the wind's path, Halladay used a rudder similar to a weather vane on a barn.

Together he and John Burnham opened a factory in South Covington, Connecticut, to make their new kind of power plant. Dr. H. A. Grant of Enfield, Connecticut, bought the first one produced and was delighted with it, but Burnham could not persuade enough conservative Connecticut Yankees to buy this kind of windmill to keep the factory going. Most investors in engines were too intrigued by steam power to be interested. So Burnham took off for Chicago, which was already reputed to be a windy city both figuratively and literally (although government records of the average velocity of the wind there have since shown it to be less than in Boston).

Railroad companies were then hurrying to lay track around Chicago. With other Chicago entrepreneurs Burnham organized the United States Wind Engine and Pump Company to sell windmills to the railroads to fill trackside water tanks. This saved the little company in Connecticut. When it could no longer deliver windmills as fast as the new company could sell them, the factory was moved to Batavia, a small town a few miles west of Chicago.

There the United States Wind Engine and Pump Company built "sectional" wind wheels up to sixty feet in diameter, and filled a colorful catalog with testimonials from happy customers. It opened branch offices in Boston, Omaha, and Fort Worth, and became the world's leading manufacturer of windmills. Five rivals also opened plants in Batavia, and for a time it was known as the nation's "windmill city," just as Detroit later became celebrated as America's "auto city."

The Halladay Standard windmill was manufactured in the United States until 1929, and for several years, under a special license, at a factory in Hamburg, Germany. A formidable competitor, meanwhile, produced and marketed a somewhat simpler kind of new windmill in Wisconsin.

### WHEELER'S SOLID WHEEL

The Reverend Leonard R. Wheeler was in charge of a Protestant mission to the Ojibway Indians on the shores of Chequamagon Bay in southern Wisconsin when he built his first windmill. Wheeler was as keen an observer of wind-driven wheels as Halladay was of steam engines. He was a conscientious family man with two sons and a daughter, and moved to Beloit so that his children could go to the new college there.

Wheeler's first windmill was a dangerous thing to erect and manage. The four wide wooden paddles in the wheel struck and injured several workmen. The dismayed minister then saw to it that a bandage was stuck on every part of a paddle that hit a man.

Neither the Reverend Wheeler's zeal to lighten men's work nor his faith in his own ingenuity ever wavered. He went on experimenting until he was able to patent a machine in 1867 that had a "solid" rather than a "sectional" fan for the wind to whirl, but that regulated itself to his satisfaction. With his son William he then started a factory to manufacture windmills, and became one of the town's leading businessmen.

The Eclipse windmill the Wheelers built in Beloit had many more and lighter blades in its rotor than the one the minister had put on his first machine. They were firmly attached to the hub and kept in the wind's path by a rudder when it was mild. But when the wind became wild, its pressure on a small vane on the side of the tower swung the entire fan wheel edgewise to the wind. That vane was attached to a weight so that when the wind's pressure lessened the "solid" rotor would be pulled back into the wind's path.

The Eclipse was a simpler mechanism than Halladay's because it did not require so many joints. Other builders found other ways to make a windmill behave itself, but a side vane like Wheeler's turned out to be adequate much of the time. Buyers could choose soon from many makes of side-vane windmills. In a technical survey of the industry in 1885, Alfred R. Wolff of the *American Machinist* recommended the side-vane windmill that A. J. Corcoran built in New York as the best one available, but he conceded that the Eclipse windmills might appeal more strongly to a "cheaper class of trade."

This may have been partly because Wheeler apparently was a better showman than A. J. Corcoran. The weights of the New Yorker's windmills were simply rectangular blocks of metal stamped AJC, but those on the Eclipses were shaped like a crescent moon. At first the weight was hung so that it resembled the symbol in the farmers' almanacs for dry weather. Later it was turned over to look like the moon symbol that promised precipitation.

Another factor in the greater success of Wheeler's factory was that a fellow churchman of the Reverend's heard about it and began to sell Eclipses in Chicago. By then Chicago had become the city where the action was, and that pious promoter of new machinery, Charles R. Morse, sold more Eclipses than Corcoran could sell of his machines.

Morse was already the western salesman for the Fairbanks brothers' scales, and also sold the Remington typewriter, a "money drawer" for storekeepers and a fare-registering device for trolley cars that preceded today's cash registers and taxi meters. The day after the great Chicago Fire in 1871, Morse rented an empty store and reopened an office in it for Fairbanks, Morse and Company with only a kitchen table and chairs. When he found that the Reverend Wheeler shared his enthusiasm for producing and peddling labor-saving devices, as well as for spreading the gospel, Morse bought stock in the Beloit windmill factory and went to work for it.

Like John Burnham, who sold seventy giant Halladays to the Union Pacific for the first transcontinental railroad line, Morse went after the railroad builders' business. He could not offer them as big a machine as the United States Wind Engine and Pump Company could, but in time Eclipses lined the tracks of the Burlington, the Northwestern, the Illinois Central and other great railroads including the Atchison, Topeka, and the Santa Fe.

In 1876, at the Philadelphia Centennial Exposition, both Halladay's and Wheeler's inventions won gold medals. From then on the factories at Batavia and Beloit competed fiercely, though there were so many other makes of windmills for buyers to choose among that neither of the two big companies ever completely dominated the field.

A hundred years after the Philadelphia Centennial Exposition, a special bicentennial show in the Smithsonian Institution's huge Museum of History and Technology included an Eclipse windmill found in Texas. An agent for Fairbanks, Morse and Company had sold it in February, 1876, but this wind-catcher had not been unpacked, assembled, or erected yet when the owner loaned it to the museum. The blades in the 12-foot rotor were wood. So was the whole 14-foot-long rudder, and the heavy tower. Only the moving parts of the engine were metal. When put up for the first time, in the museum in

1976, this windmill was one of the most impressive features of an exhibit entitled "A Nation of Nations."

### BOOSTERS AND BOASTERS

Even though electrical wonders intrigued Americans increasingly in the latter half of the nineteenth century, few men supposed that a steam-powered generator would ever produce useful energy as economically as windmills that drew it from the air. Many manufacturers filled their catalogs with glowing testimonials from owners of windmills. One wealthy gentleman, for example, reported that his windmill was the only servant he had that needed no fuel, coaxing, or driving.

The *American Machinist*'s highly respected commentator on technological progress, Mr. Wolff, also belittled the limitations on the windmill's usefulness. Most kinds of work, he pointed out, could easily be done as intermittently as the wind happened to blow at acceptable velocities. Moses G. Farmer had shown in the 1860s that the wind could even generate electricity. Three little fans drove a magneto to produce enough current to light an incandescent bulb on the little model Farmer built to obtain a patent. That model is in the Smithsonian Institution now, and still works, but Farmer's invention, like many others, was premature.

People wanted running water in their homes more than they craved a few watts of electricity then. While seeking votes in North Carolina many years ago, Senator Sam Ervin remarked to an old codger that he must have seen many changes during his long life. "Yes," the old man replied, "and they were all for the worse." Nevertheless, most families were eager to dispense with outhouses. A western writer, Francis Gilmore, has told how pleased his parents were when a windmill was installed to pump water for indoor plumbing to a tank on the second floor of their house in Idaho. William Morris, a retired New York photographer, remembers how happily his father rented a bit of land, near his home in a little Texas town, to the railroad for a windmill. Although it served the locomotives pausing there rather than the Morrises, it raised the whole family's social status. In 1976 Jimmy Carter's presidential campaign biography noted that a windmill provided the water for the first indoor toilet in his childhood home.

In 1894 the United States Wind Engine and Pump Company's handsome catalog showed pictures of windmills running pumps beside many prominent citizens' homes as well as for cattle ranches, salt mines in Texas, and an ostrich farm in South Africa. In New York City, at Broadway and Cortlandt Streets, a windmill on the roof of the six-story Benedict building pumped water to its upper floors for fire protection.

In small western towns, windmills helped volunteer firemen keep flames from destroying flimsy new wooden buildings. James W. Hill, a St. Louis engineer, cited two examples in Kansas in the *Association of Engineering Societies Journal* (Volume 3, 1873–1874): At McPherson, a windmill that could provide 30,000 to 35,000 gallons of water a day enabled volunteer firemen to squirt 2-inch jets of water at the second floor of a frame building. And at Arkansas, Kansas, another windmill kept water flowing to five 3-inch hydrants through 2,600 feet of pipes.

In New England, windmills filled big tanks for the Talbot Memorial Hall at North Billerica, Massachusetts, the Rhode Island State Sanatorium at Wallum Lake, and the Connecticut Agricultural College at Storrs. In Minnesota, Governor W. R. Marshall's mansion was provided with water for fire protection by a windmill, and in the Dakota Territory, the United States government had windmills put up at Indian forts and at the penitentiaries in Sioux Falls and Bismarck.

What better way was there for a man to proclaim his prosperity to his neighbors while making his home both more secure and comfortable? Two of Boston's leading engineering firms issued circulars describing and showing photographs of windmills that they had erected for such eminent citizens as the Honorable Robert Treat Paine at Waltham, J. B. Shurtleff, Esq., at Revere, and Howard Marston, Esq., at North Billerica, Massachusetts.

The fans of the windmills serving private houses were usually only from 10 to 16 feet wide, but Charles S. Mellen, the president of the New Haven Railroad, had a 20-foot Halladay on his estate near Stockbridge, Massachusetts. An Eclipse on the Lawrence Estate near Fitchburg was made even more impressive by an ornamental stairway that spiraled around its 70-foot tower. Visitors were invited to ascend it to admire both the countryside and the fan's responsiveness to the wind's whims.

Even in those days, it was hard to outdo rich Texans; in Fort Worth, E. E. Chase had a windmill perched on the peak of the tower of the castle he occupied.

Fairbanks, Morse and Company contended that the solid wheels of the windmills it sold could pump water in a lighter breeze than a sectional fan. A table in its 1884–1885 catalogue impressed engineers with these numbers:

|  | ECLIPSE | SECTIONAL |
|---|---|---|
| Wind receiving and working surface |  |  |
| in wind | 352 sq. ft. | 308 sq. ft. |
| out of wind | 9½ " " | 48 " " |
| Safe revolutions per minute | 50 | 25 |
| Pivots or joints in wheel | none | 92 |

The United States Wind Engine and Pump Company retorted that a sectional wheel was safer in storms and more durable than other fans. But by 1893 it was making and selling three other types of machines in addition to its Halladay Standard. They were a U.S. Standard Wheel, a Vaneless Standard Machine, and a little Gem Steel Wind Engine.

The Gem was said to need no oil because it had graphite bearings. This was an important advantage. Windmills cried out and balked when not kept lubricated, but oiling a windmill from a tiny platform near the top of a swaying tower was dangerous work. Many men fell and were seriously injured. One maker hung oil cans on his towers so that a person on the ground could tip them by pulling strings. Others ballyhooed towers that could be lowered, and longer lasting lubricants.

Sales of windmills to railroads, public institutions, and homeowners declined alarmingly in the first half of the twentieth century. By then the industry was indelibly stamped with the characteristics of the region that Colonel Robert R. McCormick liked to call "Chicagoland," and changing fast. The United States Wind Engine and Pump Company's adaptability enabled it to survive until 1951. Fairbanks, Morse and Company became better known for other products than for Eclipse windmills. The company that made them famous is now a subsidiary of Colt Industries, Inc., and no longer manufactures windmills within the United States.

# The Windy Wild West

I've reached the land of drouth and heat,
Where nothing grows for man to eat,
The wind that blows with burning heat
O'er all our land is hard to beat.

A PARODY MARI SANDOZ HEARD SUNG

NATURE LEFT a broad ragged ribbon of nearly smooth land in our continent which cartographers once called the Great American Desert. It ran north from Texas through the Oklahoma, Kansas, Nebraska, and Dakota Territories to Canada, and has been designated the Great Plains since fan wheels and governors for windmills were invented.

The wind roared and romped so madly there that nearly every vestige of life was once rubbed off that corridor except the tough grass the buffalo ate. Nomadic Indians followed the buffalo herds, and homeseekers whipped their horses to race across that arid part of the prairie to the green valleys in the foothills of the Rockies. Only cattlemen, horse traders, and other gamblers lingered along the Plains trails.

To everyone old enough to have seen the Great Plains a few score years ago the changes that time has brought in them now seem incredible. Certainly they were changes for the better.

From a river bluff a few miles west of Omaha the land as far as one's eyes could see—all the way to the horizon—was once as forlorn as an empty parking lot. Young brides seeing that lonely plain often urged their husbands to turn their covered wagons back toward the muddy Missouri River they had just crossed. Cowboys agreed that Nebraska was no place for a woman unless she could keep a sod house tidy, shoot a snake, and climb a windmill.

The Platte River crept across Nebraska reluctantly. In winter it was a long sheet of thin ice that melted in the spring to become a brown streak of water often a mile wide but only a foot deep. William Jennings Bryan's critics likened his oratory to it. There were few trees to provide shade in July or fuel in January along the Platte or the creeks it absorbed. Storms brewed in the Gulf of Mexico sent tornadoes swirling into this part of the continent, and Arctic gales whistled back through it. The weather changed so quickly and often that people who griped about it were told just to wait a minute.

The noted historian of the frontier, Walter Prescott Webb, reported that three inventions were essential to settlers' survival on the Great Plains. They were the Colt revolver, barbed wire, and cheap windmills.

Samuel Colt had gone to sea from Connecticut when he was thirteen, and is believed to have gotten the idea for a six-shooter with revolving chambers from the pilotwheels on sailing ships. Joseph Farwell Glidden of Illinois made barbed wire by having a boy climb a windmill tower. The lad put short sharp barbs on the upper end of a greased wire, and when they slid down it Glidden pulled them a short distance apart and twisted the wire to hold them in place. Revolvers disposed of the west's badmen and barbed wire kept cattle off farmers' fields.

Neither guns nor fences could subdue the west's wild winds. People said they were more powerful than Niagara Falls and eagerly tapped the wind to moisten the arid land. Windmills usually appeared first in forks between the creeks where water could be pumped from shallow wells for herds of cattle. When the government allotted square bits of the open land to homesteaders, barbed wire kept the cattle off the fields the claimants plowed. As those settlers advanced into higher, drier parts of the Great Plains wells had to be deeper and windmills more powerful.

Homesteaders in that hostile environment found themselves, in Professor Webb's memorable words, "far from markets, burned by drought, beaten by hail, withered by hot winds, eaten out by grasshoppers, exploited by capitalists, and cozened by politicians." Windmills, indifferent to men's cupidity, pumped water to bolster the farmers' hopes, and relieved life's monotony for children in lonely hovels. Willa Cather and other famous children of the frontier have mentioned windmills fondly. Girls cupped their hands to drink the cool water gushing from pumps below the towers. On an outhouse wall behind a one-room school in Cherry County, Nebraska, one child scribbled:

> We like it in the sandhills,
> We like it very good,
> For the wind it pumps our water,
> And the cows they chop our wood.

### MULTI-PURPOSE MACHINES

John Wise, in a memoir of his life on the Plains, recalled how tired his mother became pumping water for livestock, and how grateful the family was to a neighbor who loaned his father a hundred dollars to buy a windmill. In Kansas "a windmill and a pump" on every farm without a brook was a popular slogan, and Everett Dick, in *The Sodhouse Frontier*, told of having counted thirty-nine windmills when he awoke in a Kansas camp one morning.

Sweating men came in from the fields, chased the cows away from barnyard tanks, and plunged into them. Prairie preachers dipped babies and repentant sinners in those tanks to save them; one of John Stewart Curry's most famous paintings, now in the Whitney Museum in New York City, depicts a baptism at the foot of a windmill tower.

The water the windmills pumped both cleaned the family's clothes and kept their food cool. Autumn Ryan, whose son is now a Congressman from California, enjoys describing the tidy washhouse that her father built in the base of his windmill in Oklahoma. He painted it white to match their new frame farmhouse, and her mother kept baby chicks in the washhouse until a weasel chewed its way through the floor boards. When the windmill screamed angrily Autumn's dad rushed the family into a storm cellar, where he once stood in the doorway to watch the funnels of seven tornadoes devastate his neighbors' fields. When the windmill was silent he sometimes let Autumn climb its tower to pull out a pin so it could purr like a kitten.

"A stranger journeying by rail in northwestern Kansas," *The Kansas*

*City Star* reported in 1904, "is struck by the many windmills he sees from the car windows. The prairie land is fairly alive with them. The windmill has taken the place of the old town pump, and no western town is complete in its public comforts without a mill supplying water to man and beast by the energy of the wind. In the town of Colby, Kansas, on the Chicago, Rock Island, and Pacific railway, 125 private windmills bring water to the top for domestic consumption. The home of the humblest citizen of the town is not complete without its windmill."

Among other things those windmills watered vegetable gardens and fruit-tree seedlings. This so improved the pioneers' diets that one windmill admirer predicted the people would some day be healthy enough that they would stop buying the patent medicines peddled everywhere by itinerant "doctors."

Children stopped playing in a windmill tower after a posse hung a man from it, but that happened in real life much less often than in the imaginations of pulp magazine writers. In 1964, when electric pumps had replaced most of the windmills in the no longer wild west, another *Kansas City Star* reporter wrote affectionately about the windmill that stood in his father's corral when he was a boy. He still remembered that it "had all the usual sounds—the gentle sough of the breeze through the fan, the creak of the tower and the rhythmic metallic working of the gears and sucker rod, and finally the steady soft pouring of water into the tank.

"And then," he continued, "there was a sound all its own: A blade bent by some long forgotten encounter with the wooden tower at each revolution—thump, thump, thump, when the wind was soft and at a machine-gun rattle when a summer storm sent up dust and dead tumble weeds racing across the flat to tangle together at the first fence row. . . .

"In a storm at night, when thunder crashed and in the lightning flashes we could see the cottonwoods whipping frantically and hear the frightening snap of the limbs, the windwheel whipped back and forth as if it could not decide in which direction to face the gale. On quiet nights, for a child awakening in the darkness of a stuffy room, the chipping of the wounded blade against the tower took on a comforting sound, reassuring that this was home, that there was no storm, and that the windmill was pumping water."

103 ह‍

How sad it is, many grandparents sometimes think, that today's children lack such comforts.

## WOOD VS. METAL, ETC.

The prairie windmills seemed as alike as Indians' totem poles to some strangers crossing the Great Plains in the latter half of the nineteenth century. But the men who built and bought those mills found the variety confusing. Small factories exploited inventors' patents on mechanisms to improve their windmills. Salesmen touted gimmicks and made extravagant claims for both significant and trivial innovations. Buyers' assessments of technological developments at that time, being based mainly on hearsay, consequently differed as much as small boys' opinions of new cars differ today.

The blades in the first fan wheels manufactured in America were wood. When some companies began to make metal fans, many farmers wondered how much the builders knew about the differences between wood and metal. The farmers knew that the wind was more accustomed to blowing trees around than to powering metal machinery. Some feared that metal blades in windmill fans would "rust the wind" that struck them.

Mast Foos and Company in Springfield, Ohio, was probably the first of several companies that produced iron windmill rotors. It advertised a turbine patented in 1872 which had seven iron buckets providing "a combination of curved and spiral surfaces" that was said to present the best possible angle to the wind. But some people thought that those metal rotors looked too much like a child's pinwheel to be much good. In 1976 the Smithsonian Institution could not find a Mast Foos windmill anywhere, and had a new one built from drawings for a special bicentennial show.

For several years the big windmill companies offered their customers their choice of wooden or metal rotors. Galvanizing the blades prevented them from rusting. But some builders coated blades this way before assembling them and others did it afterwards. How was an inexperienced farmer to know which was the better way? And that was only one of his many dilemmas. The history of one of these metal machines, the Plymouth Iron windmill, reads like fiction from *The American Boy*. This machine was patented by Clarence J. Hamilton,

*An advertisement for Mast, Foos & Co.'s iron wind turbine.*

who fixed watches and clocks in the window of R. L. Root's jewelry store in Plymouth, Michigan. It so impressed Root and other local businessmen that they started a factory in 1882 to produce Hamilton's iron windmill. Meanwhile the inventor conceived and patented another way of using the power of air. It was a gun that quickly became so much more popular than his iron windmill that the two-story factory stopped making windmills—because every square foot of space in the plant was needed to fill the orders from small boys everywhere for Daisy Air Rifles.

Although both Halladay's centrifugal governor and Wheeler's side vane continued to be highly regarded, other ways were also found to make a windmill behave itself. Althouse, Wheeler & Company in

Waupun, Wisconsin, was one of several makers that put a rudder in front of a fan wheel rather than behind it. The wheel itself then served as a vane to swing the whole engine to the right or left when the wind veered. Other manufacturers dispensed with a side vane by mounting the rudder parallel to the main shaft but a few inches to one side of it. This kept the fan from ever fully facing the wind, but quickly swung it out of the way when a sudden gust hit the rudder.

Like inventors in England, Americans also tried to regulate wind-driven engines with springs. The Stover Manufacturing Company at Freeport, Illinois, claimed its windmill factory was the biggest one in the world. It made Ideal towers and Samson engines, and favored springs because they could act faster than other mechanisms. Competitors who used suspended weights insisted that they were easier to adjust properly.

Safety was another consideration. People were injured while fixing, painting, and adjusting farm windmills. Oiling one was so dangerous that it hastened the use of longer lasting lubricants and bearings on the trembling towers. Salesmen also insisted that some makes of windmills were better than others because they had fewer nuts and bolts that might come loose. Like automobile makers later, the windmill manufacturers turned out whatever sizes and types of machines they could persuade people to buy—sometimes without testing innovations adequately before marketing them.

The Butler company in Indiana claimed in 1903 to have produced "the simplest mill in the world." The metal blades in its fans were coated with a zinc alloy to keep them from rusting. The main shaft ran on patented ball bearings that could be replaced without taking the engine down from the tower. Oil ran up a wick to cups from which it dripped onto those bearings, and the company guaranteed that its Klondike Windmill Oil would not congeal in zero weather.

Many buyers chose towers that they could erect themselves, and some makers advertised complete "do-it-yourself" water supply systems with all the latest accessories. Smith and Pomeroy in Kalamazoo, Michigan, and the Flint and Walling Manufacturing Company in Kendallville, Indiana, produced kits like these, including regulators to stop a windmill when a water tank was full. Instead of a tower, the Kalamazoo company would send seasoned pine to men who wanted to build their own support for its Eureka Wind-Mill. And the Hoosier

Pneumatic Supply System that Flint and Walling advertised included "No-Oil-Em" bearings for its Star windmill.

### F.O.B. CHICAGO

Sears, Roebuck & Company offered "the most complete line" of windmills on the market. It included "direct-stroke wooden," "direct-stroke steel," "back-geared pumping windmills," and "high-geared steel power mills." Buyers could choose any of several sizes of engines as well as types and sizes of towers. The most popular machines had 8-foot rotors and towers 30 or 40 feet high, but Sears recommended a larger "suburban" model to people who wanted running water in their homes. A bright red border around the rudder and the tips of the blades indicated that Sears' machines were all Kenwood brand windmills.

Many manufacturers painted their names in huge letters on the rudders of their windmills. As a special inducement to buy a Kenwood, Sears offered to put each customer's own name on the rudder, "free of charge." Those rudders were such prominent billboards that advertisers sometimes offered to pay farmers to let them paint trade marks and slogans over the original labels. If windmills were as widespread today as they were seventy-five years ago, you might see "Geritol, Every Day" on a rudder beside a farmhouse as well as on your television screen.

A special 118-page supplement to Sears Roebuck's fat catalog of other merchandise cited the advantages of each type of wind-driven engine, tower and pump that the company sold. It also included detailed directions for choosing a windy site and erecting a windmill. Chicago's other big mail order house, Montgomery Ward, sold similar machines at competitive prices. A farmer who liked to do everything possible himself—and knew how—could buy a small wind-driven engine for less than twenty dollars and build his own tower for it over his old well pump.

But a man usually needed someone's help when several hundred pounds of metal parts arrived from Chicago for him at the nearest depot. Dowsers became experts in putting up towers over wells where their forked sticks had told farmers to dig for water. Idle drifters on the Great Plains became specialists in adjusting and repairing the intriguing new machines. One man had twelve children while helping to put some four thousand towers over wells in Oklahoma and Kansas, and taught the tricks of his trade to all twelve so that they could help him.

Cowboys found it easier to spot stray cattle from the platform near the top of a windmill tower. Many soon learned to climb a wobbly framework with a brown beer bottle full of oil in one hand to pour on squeaking gears. If a man did this before the machine's screams drove a homesteader's wife batty, she might cook something extra special for him in the sod house that she fought to keep clean.

For more than fifty years, Guy Skiles built and fixed windmills around Pumpville, Texas, where a big one pumped water for the Southern Pacific's first locomotives. Skiles later put one up to pump water from a spring two hundred yards away to the swimming pool at his son's home. "There's something fascinating about a windmill, like watching a fire," he explained. But he added that it was hard to feel romantic about a windmill when you were up on an icy platform trying to fix it on a windy winter day.

Even the catalogs of many American windmill factories' products are hard to find now. Historical and other libraries have saved some, fortunately; many are fun to examine, especially if you like puzzling pictures. Engineers drew some of the illustrations to get patents, and so cluttered them with tiny numbers and letters that it is hard to figure out what each part did. The texts purporting to explain those machines' merits are sometimes as implausible as many modern television commercials—but the letters from customers quoted in the catalogs make it perfectly clear that they loved their windmills.

The weights used on many makes of windmill to keep the fans swung in and out of the wind's way are highly prized by collectors today. Often the weights were shaped to resemble squirrels, roosters, camels, buffalo, or other symbolic creatures. Homer Sargent, of York, Nebraska, began collecting windmill weights several years ago; by 1975 he had eighty different kinds and hoped to find forty-five others that he had heard about. They ranged in weight from 8 to 135 pounds, and some were hollow so that their weight could be changed by putting buckshot inside of them or dumping some out. One Nebraska factory used weights shaped like a fat bull. A competing

*In a photo from about 1868, a large windmill pumps water to fill a tank on the line of the Union Pacific Railroad near Laramie, Wyoming.*

company in a nearby town chose horses; long-tailed horses then could be put on this firm's big windmills and short-tailed horses on its little ones.

Nick Arendt, a farmer near Belgium, Wisconsin, still had a shop-made windmill in 1968 that he traded forty-five dollars' worth of lumber for sixty-five years ago. When a *Madison Capital Times* reporter asked him about it, he reported he had spent only $14.50 for new parts for it. "A gallon of oil lubricates it for a year," he explained. "Doesn't need any more attention. Durn good machine."

Farther west in what was once a desert, an admirer of Americana may still find many windmills like that one. After driving across the Great Plains in 1975, Eugene L. Meyer reported in *The Washington Post:* "West of Oklahoma City, the land opens up with fewer trees and occasional strange-looking mounds of earth. There are a few oil wells, but mostly it is ranching country, with windmills dotting the landscape. . . . In the Texas panhandle, there appears in the middle of the treeless plain a modern dome house. It seems a logical successor to the sodhouses of the early settlers in this lumberless land . . . on the 1930s Okie Route 66, a patch of sand billows up as if on cue. This occurs near McLean . . . where you can buy a windmill or have one fixed."

X

# The Industry's Fate

The winds must come from somewhere when they blow,
There must be reasons why the leaves decay;
Time will say nothing but I told you so.

W. H. AUDEN*

NO ONE DOUBTED a century ago that the wind was the cheapest source of power known. Wherever it came from there was always more of it, and country folk knew where it was strongest. Before utility companies lured us into paying them every month for electrical energy, many Americans enjoyed having an inexpensive little power plant so simple that a child could understand and learn to fix it.

Why then did nearly all of our first windmill factories vanish? One reason may be that, like the hare who lost a race to a plodding tortoise in an old fable, most of the manufacturers dozed too long. Few paid attention to scientific research that John Smeaton in England and Charles Coulomb in France had begun, or to P. LaCour's wind-driven generators in Denmark. But the United States Wind Engine and Pump Co., hired Thomas O. Perry, a professional engineer, to evaluate its own and other manufacturers' machines.

Perry tested sails for windmills on a centrifuge that he ran with a steam engine in a sealed laboratory. He could control that engine's speed, and mount rotors for windmills on the long arm of his centri-

---

* "If I Could Tell You," from *Collected Shorter Poems* 1927–1957, by W. H. Auden, copyright, published by Random House, Inc.

fuge to whirl them around in still air. By thus simulating the impact of wind on a rotor, he found the velocities needed to energize 61 different kinds of wind wheels. That was the sort of data aeronautical engineers now get from wind tunnels, and was so useful to the company that it kept Perry's findings a trade secret until 1889.

Fortunately, however, no company could seal the lips of other researchers like Professor F. H. King at the Agricultural Experiment Station in Wisconsin. Bewildered buyers wanted to know what they could expect from each one of the many makes, sizes, and types of windmills being manufactured, and soon found out. In Kansas where many varieties were running pumps, Edward Charles Murphy measured their horsepower in 1895 and 1896. In his voluminous report of this field work, Murphy named the makers and included the specifications of the windmills he studied. The United States Geological Survey published that report in 1901. This was a noteworthy pioneering effort by a government bureaucracy in consumer education.

Near Garden, Kansas, where Murphy did his research, wells were usually square holes 3–4 feet wide, cased with wooden siding, and from 8–40 feet deep. The windmills above those wells drew water from sand and gravel beds either to fill tanks for livestock or to irrigate up to 15 acres of land. Some were "pumping" mills that drove a rod up and down, and others were "power" mills in which vertical shafts revolved. Some stroked a pump every time the wind blew the rotor around, and some were "geared back" so that the fan would go around several times to stroke the pump once. The pumpers were generally used to water cattle, the "power" mills for irrigation.

To determine the horsepower obtained from each windmill, Murphy noted the size and lift of the pumps, the useful work done per stroke, and the number of strokes per minute when the wind blew 8, 12, 16, 20, 25 and 30 m.p.h. The table in which he presented his data filled a page of small type when the Government Printing Office published it. There were blank spaces in some columns, and neither the original cost of the machines nor the cost of erecting and maintaining them was included. But many numbers in that tabulation were enlightening both to manufacturers and consumers.

No rig was delivering more than one useful horsepower at wind velocities less than 25 m.p.h. Even when the wind was strong, most windmills were using only a small fraction of a horsepower from it to pump water. The big windmills, of course, were generally stronger than the small ones—and a metal fan was preferable to one with wooden blades. Here are a few numbers from Murphy's table:

| WINDMILL | WHEEL DIAMETER | USEFUL HORSEPOWER AT WIND SPEED OF 25 M.P.H. |
|---|---|---|
| Halladay | 30 ft. | 1.070 |
| Eclipse | 22.5 | 0.182 |
| Jumbo | 15.5 | 0.154 |
| Aermotor | 16 | 0.601 |

The Halladay and the Eclipse came from big factories, and the Jumbo was a homemade windmill. All three had wooden sails. The Aermotor was manufactured by a new factory, and was a direct result of Thomas O. Perry's research with metal sails.

### THE MATHEMATICAL MILL

Incandescent electric lights began to brighten American life while Perry was working for the United States Wind Engine and Pump Company, but that did not worry the windmill industry. Sir William Thompson, the eminent British scientist who showed how Niagara Falls could be made to generate electricity, thought that the wind's power could be used similarly. And so did many other men, as we shall see.

In 1883 Perry had patented a metal windmill fan. When no longer employed by the United States Wind Engine and Pump Company, he went to work for LaVerne W. Noyes, who was trying to develop a self-binding threshing machine in Chicago. Noyes was also interested in windmills, and encouraged Perry to keep working on them. Perry then designed an all-metal machine with curved sheet steel blades. They were set at a pitch to catch as much power as seemed possible from the wind, and the rotor was back-geared to run a pump smoothly.

Noyes galvanized the metal both to protect it and to improve the looks of the wind wheel. He then found ways to manufacture this machine economically, and put the Aermotor on the market. He called it "the first truly scientific windmill," and competitors derided it as "a mathematical mill."

The first Aermotor towers that Noyes built each had five glass cups on them which a milltender could watch to know when the gears needed more lubrication. He could then refill the cups without climbing the tower, by tilting it like a bottle on a bar. Such service was no longer needed after Daniel R. Scholes developed an automatic oiling system for the Aermotor in 1915. From then on other manufacturers tried harder to improve their products, too, though the demand for windmills was shrinking.

The Wright brothers' flight had focussed the attention of men interested in aerodynamics on propellers for airplanes rather than impellers for pumps. Flying machines were essential both to the nation's defense and its prestige, and the United States government provided the funds for the National Advisory Committee on Aeronautics (NACA) to establish a great center for aeronautical research at Langley, Virginia. This happened between the two world wars, and the Langley Research Laboratory has both contributed greatly to aeronautical progress and provided leaders for the big centers our space agency now has elsewhere.

In 1928, NACA published a Technical Memorandum by Professor A. Betz, who had found that at best a wind-driven wheel could only extract about 59 per cent of the wind's energy from it. In that memorandum, Betz pointed out that windmills already were highly developed and expressed doubt that they would be improved greatly by further scientific research. Fortunately, some engineers were not discouraged by Betz's "law," and we now have more efficient machines than we had in the 1920s.

According to Jim Fetters of the Aermotor Division of Braden Industries, nearly a hundred windmill factories were still competing in 1929 for sales totalling about $10 million. Some had begun building windmills to generate electricity, which was much more difficult than producing machines to run pumps by mechanical energy drawn from the wind. But the industry did not come through the Second World War with the vigor that it still had after the first one.

Aermotors were still being built in Broken Arrow, Oklahoma, until recently, but that factory has been moved to South America. On a Children's Farm maintained by the National Park Service at Oxon Hill, Maryland, you can still see an Aermotor engine pump water into a barnyard tank for the farm animals. The old gray tower on which the motor is mounted, however, was built in Pennsylvania. Youngsters were leaping out of cars to watch that windmill the Sunday I visited the Children's Farm, but none of the attendants realized that the water was being pumped by an engine as historic as the Aermotor.

By 1973 only two factories in the United States were still well equipped to make complete old-fashioned farm windmills. Neither one of those factories was noted for its owners' interest in sponsoring scientific research, and only one of them still bore its founder's name. Most of the factories that once built the American type of windmill had gone out of business or were manufacturing something else instead of windmills.

## OUR OLDEST FIRM

Statistics have shown that people live longer in southeastern Nebraska than in most places, but no one is sure why. Windmill factories have thrived there, too. The Dempster Engineering Works in Beatrice became ninety-eight years old in 1976, and a rival, the Fairbury Windmill Company in the next county seat town, vied with it for business until 1961.

Beatrice is on the Blue River, a steady enough stream to provide a bit of water power. Its principal attraction was a land office when Charles B. Dempster opened a shop in the little frontier town's dusty main street. He was then twenty-five years old and had only thirty dollars. His brother and a friend loaned him $300 to go into business. Homesteaders were claiming a million acres of land at Beatrice, and "C. B." Dempster sold farm equipment to them in his tiny shop.

So many customers wanted windmills that Dempster began to manufacture them. By 1885 business was so brisk that he built a two-story factory on the riverbank. During the hard times later, he drove around the state in a farm wagon full of parts for windmills. Wherever he found a farmer struggling along without one, "C. B." was ready to help him put one up. If a man did not have enough money to pay for a windmill, Dempster would accept eggs, hay, or what-have-you. When he was tempted to close the little factory on the riverbank and move his business to Omaha, his neighbors bought and gave him a site for a new and bigger plant in Beatrice.

Business boomed again in the 1900s. The Dempster company's

sales then exceeded a million dollars a year. It had branch offices in Omaha, Kansas City, Des Moines, Sioux Falls, Amarillo, San Antonio, and Oklahoma City. When sales in the United States began to decline, R. E. Lindell found big markets for Dempster's windmills in Central and South America. The Beatrice *Sun* often told its readers that the string of freight cars on the town's Burlington Railroad siding was being loaded with windmills for shipment overseas. The Dempster Manufacturing Company survived the Great Depression in 1929 by making pumps, cultivators, and other farm implements as well as windmills. A *WPA Guidebook* published in 1930 listed the factory as a point of special interest in Nebraska, and people went out of their way to be escorted through it.

Charles B. Dempster's son, Clyde, succeeded him as head of the business, and ran it until shortly before he died in 1974. The biggest windmill fan that the company ever produced was only eighteen feet wide, and when farmers began to buy electric motors the Beatrice factory built electric pumps, fertilizer applicators, and other fancy new farm equipment. But its catalog continued to show both stubby little pumps with spouts like those that water gushed from long ago in many kitchen sinks and big barnyard pumps with "extra long handles" for use when the wind failed to blow.

In Fairbury, a few miles west, people were as proud of their town's windmill factory as the citizens of Beatrice were of the Dempster plant. Both companies produced metal machines that withstood the prairie fires started by sparks from steam locomotives. Later, when dust storms drove the Okies to California and farmers in the Dakotas to Oregon, newsreels of that grim exodus from the Great Plains showed windmills built in Nebraska still standing. Workmen in the Fairbury movie theater assured each other that those storms never blew down one of their windmills, and applauded the dismal films.

During and after World War II, servicemen from Fairbury reported seeing windmills from their home town on other continents. Tourists, too, told of seeing them from Korea to Timbuctu. Morris Speir, the factory's manager, bragged for years that "the sun never sets on a Fairbury windmill." It may not have yet, though the Fairbury Pipe and Supply Company stopped making windmills for keeps in the 1960s.

A century ago the Annu-Oiled Windmill still pictured today in the loose-leaf catalog of Dempster Industries, Inc., would have made many a farmer's mouth water. A spider wheel holds fifteen heavy curved steel blades tightly in place to intercept the wind. A large galvanized steel vane that cannot sag or buckle guides the wheel. A double spring bumper absorbs the shock when the fan is swung into the wind on a ball bearing turntable, and tapered roller bearings absorb the thrust and wear on the main shaft assembly. The teeth in a box full of gears are machine cut and meet in an oil bath that needs to be changed only once a year, and a galvanized steel hood fits snugly over the working parts.

The fan may be from 6 to 14 feet wide, and is mounted on a tower 22 to 61 feet high with legs from 5 to 16 feet apart at the base. Nearly everything is metal except the little platform near the top and a long rod that runs down to the pump. Steel guides keep the pump rod in line, and a regulator prevents the water tank from flowing over.

When I drove from Omaha to Beatrice in the summer of 1973, I saw many windmills still standing near the farmhouses, but none still running. Near Beatrice most of them had been neglected as much as anywhere else. I found the Dempster plant was a large two-story conspicuously modern factory building on the main highway through the town. By then a group of local businessmen had acquired the company from Clyde B. Dempster, and there was not even a picture or model of a windmill in the attractive reception room for customers.

The pretty girl who greeted me looked surprised when I asked to see a windmill, made several phone calls, and finally directed me to the office of R. E. Lindell, an elderly vice president who sold windmills overseas for many years. He enjoyed reminiscing about the days when nearly every caller wanted a windmill, assured me that the company could still produce one if necessary, and was sorry that it did not have one in stock.

Lindell politely pointed out that a wind-driven pump would cost me five or six times as much as an electric pump that the company could deliver immediately. He reminded me that oil was found in Nebraska in this century, that the rivers had been dammed to store water for irrigation, and that there were fewer places now where a transmission line did not deliver adequate power for electric motors. After the terrible dust storms of earlier years the government had planted a shelter belt of trees to retard the wind on the Great Plains, and had done nothing since to encourage the use of the wind's energy.

When enough orders piled up, the Beatrice factory would probably produce another batch of windmills to fill them, mainly for the sake of its reputation. As an officer of the company Lindell was currently more concerned about the rising cost of steel than environmental pollution or an energy shortage. While I thought about this wise and philosophical gentleman's remarks about windpower on my way back to Omaha with a strong fresh wind in my face, a nursery rhyme by Robert Louis Stevenson popped back into my mind:

> I have a little shadow that goes in and out with me,
> What can be the use of him is more than I can see.

### BAKER'S SIMPLE WINDMILL

By the 1970s the leading manufacturer of the American multi-bladed type of windmill was the Heller-Aller Company on the Maumee River bank in Napoleon, Ohio. It was only mildly afflicted by the "growthmania" fatal to many businesses. Like the proverbial shoemaker, it had "stuck to its last," and its president, William J. Selhorst, was still exuberantly selling a remarkably simple machine to pump water.

Frederick Baker, a local blacksmith, designed and built a windmill in 1883 that soon became quite popular. Three years later a company was formed to manufacture it, and in 1889 the firm moved into the 40,000-square-foot factory that it still occupies. Baker's first wind wheel had wooden slats, but the new factory was especially designed to produce galvanized metal windmills, tanks, and regulators.

The Heller-Aller company was named for the two men who built towers for Baker's engines. A crank on the shaft turned by the fan raised and lowered a long rod attached to a pump every time the wind blew the fan around. The mechanism would fit on almost any kind of tower, and selling towers for radio and television stations has helped the firm survive.

Selhorst was a farm equipment salesman when he bought an interest in the Heller-Aller business several years ago. Business was so bad in the 1960s that Selhorst was tempted to sell its patterns, jigs, and tools and close the old factory. For a while the company sold only about a hundred windmills a year. But Selhorst kept it going like a storekeeper in a dying town because it was "a pretty good way to make a living," and he was convinced some people would always want windmills.

In the heyday of the industry the company developed two versions of the so-called Baker Direct Stroke Steel Mills that were said to last a lifetime. It called its large machines Sailors and its small ones Pirates, and sold them in nearly every country on earth. To make them responsive to the wind it put ball bearings in the turntables on top of the towers, and enclosed the moving parts so that they would dip into oil to lubricate themselves.

The fans range from 6 to 12 feet in diameter and contain more blades than some other makes of windmills. The 8-foot wheel, for example, has 6 sections with 6 slightly concave galvanized steel sails in each one. Both Baker "run-in-oil" and Dempster "annu-oiled" windmills run quietly and smoothly, and even in a light breeze their simple direct-stroke windmill can pump enough water to satisfy many farmers, but Selhorst recommends his back-geared machine to those who need enough power for deep wells.

Selhorst's most faithful customers have long been members of Mennonite colonies, who shun the use of electricity. There are tens of thousands of such "plain people" in Amish and other communities in Pennsylvania, Maryland, Ohio, Indiana, Wisconsin, Missouri, Iowa, and Canada. Many members of their sects still do not have telephones and prefer horses and buggies to automobiles. For such people the wind has pumped water since the sixteenth century.

West of the Mississippi River, the Heller-Aller Company is fairly well represented by farm equipment distributors, but in the east most of its sales are made direct from the factory without a middleman. In 1972 the company produced and sold about five hundred windmills, and Selhorst is optimistic about his business for several reasons.

The increasing cost of raising cattle has revived ranchers' interest in the wind's power. Storm damage to transmission lines, low voltages, and high utility rates have also made electric pumps less attractive. Selhorst points out, too, that there are now more ponds and lakes on American farms in which water can be stored than there were in the good old days.

"Ecology buffs" have further increased the demand for simple windmills. Many well-to-do gentlemen who have bought and retired to New England farms have felt sharp pangs of regret for rural life's lost simplicity. Both nostalgia and concern about pollution have prompted people to drive to Napoleon to see about buying a windmill. But Selhorst reports that a good many give up the idea when they find out what a windmill now costs.

Neither the Ohio nor the Nebraska factory has ever produced windmills to generate electricity. If that is what a customer wants, their salesmen may refer him to a firm in Sioux City, Iowa, or to some importer of machines from other countries.

# Battery Chargers

How you gonna keep 'em down on the farm
After they've seen Pa-ree?

WORLD WAR I SONG

WHEN THE YANKS came marching home in 1918, many American families still went to bed with the chickens. Delco and other tiny electric plants lit up some rural homes, and transmission lines were being extended to more, but not fast enough to satisfy high-spirited young folk in the roaring twenties. You could read by an oil lamp, but you couldn't have a wireless receiving set without a bit of electricity.

In the nineties, an electric bulb had glowed on current generated by the wind for Fridtjof Nansen while he searched for the North Pole in a long Arctic night. *The Electrical Experimenter* had told its readers in 1915 that electric lights from a windmill could be very economical at an installation cost from $75 to $100. That was a lot of money, but there were lots of experimenters, too.

Some were war veterans wearing out khaki pants, and some were small boys in knickerbockers. They could fill out coupons in magazine ads to buy parts for homemade wireless sets. When you covered a board with coils and junctions, clamped an earphone on your head, and tickled a cheap crystal just right with the tip of a fine wire, you could sometimes hear the faraway chatter of a telegraph key. Later on, when you might hear some distant person talking and even singing, grandparents began to stay up nights with the youngsters to take turns listening.

A small battery could provide enough current, and a windmill could keep it charged. "We would take a generator from a Model T, rewind it, hook it to a windmill, and use it to charge our radio batteries," a physician in Lawrence, Kansas, recalls. The windmill might be made with an old propellor from a Curtiss Jenney. When anchored to a barn roof, the prop's two blades revolved faster than the fans in barnyard windmills. This was not what Ford generators or airplane propellers had been designed to do, but who cared? The rotor's speed was what mattered.

Men who could not find old propellers carved imitations. In North Dakota, where the wind blew eight hours a day, so many people wanted small aerogenerators that the agricultural school's faculty at Fargo was besieged with requests for advice about making "impellers." Many do-it-yourselfers hoped to obtain enough current from the wind for a few light bulbs as well as a radio, and some did.

*North Dakota Agricultural Experiment Station Circular 58,* entitled "Homemade Six-Volt Wind-Electric Plants," was first issued in 1935, revised and reprinted the next year, and again in 1939. It warned amateur builders that making an airfoil to intercept the wind satisfactorily would be "a tedious task requiring patience," but told them how to go about it. The authors suggested that one begin with a board 5 feet long, 5½ inches wide, and nearly a full inch thick when planed. Straight-grained soft pine could be used if there were no knots or cracks in it, but a straight piece of hardwood would be better.

This two-bladed impeller and a generator could both be mounted on one end of a horizontal board with a vane attached to the opposite end; the board was pivoted on top of a vertical iron pipe. Detailed directions were included for doing everything, including choosing a generator, wiring the plant, and caring for storage batteries. A table suggested that a 6-volt plant might keep storage batteries sufficiently charged to light two 50-watt bulbs for up to 15 hours.

Similar plans were published and circulated in other states. With a little bit of luck a man might not have to spend more than five or ten dollars for parts to build an aerogenerator. No one knows how many amateurs tried or how much current their machines delivered, but every few watts whetted farmers' appetites for more. Some people asked workmen building transmission lines how long it would take current from a central power plant to flow through the wires to their homes—and found it hard to believe that, if they measured time by the sun's position in the sky and lived several miles west of the plant, the electricity would arrive before it was generated.

Hopeful users of windpower to generate electricity soon learned that a wheel in the wind had to whirl much faster to produce electrical current than to run a pump, that the wind was seldom both strong and steady, and that wind velocities were hard to predict reliably.

Some of the big windmill companies recognized their customers' rising expectations after World War I. The Perkins Corporation at South Bend, Indiana, was one of the oldest firms, having been established in 1860. It developed the Aeroelectric farm lighting plant driven by "a scientifically designed aeroplane type of propeller" with only two blades. They weighed much less and exposed less surface to the wind than a fan with many blades. The whole mechanism was lighter, less vulnerable to storms, and generated more power from the wind than the engines designed for pumping.

The Aeroelectric was rated at one kilowatt, and was used to charge storage batteries. The engine could be mounted on a tower that a farmer might already have. "We do not hesitate to recommend this plant to any one not requiring to exceed 50,000 to 60,000 watts per month on an average the year round," the manufacturer asserted. Purdue University had one, and people with no other source of electric power testified to this plant's usefulness to them.

More aerogenerators might be running today if more people could have afforded them. Private industry began the electrification of the nation and the government completed it. Alternating current generated by big thermal plants at high voltages could be sent long distances cheaply via transmission lines. The variable speed of the wind could only be used, at the time, to generate direct current. It could be accumulated in storage batteries, but that increased the cost of home-lighting plants.

### THE JACOBS GENERATORS

In 1975 the Wind Energy Society of America, meeting at the University of Southern California, presented a Four Winds Award for Excellence to Marcellus L. Jacobs. He built aerogenerators years ago that searchers for a used one value most highly today.

Jacobs was born in North Dakota, and grew up on a Montana ranch thirty miles from the Fort Peck dam on the Missouri River. While he was still in high school, he put together "little peanut radios" to sell. Later he took a couple of short courses in electrical work in Indiana and Missouri, and learned to fly. Jacobs began experimenting with wind-driven engines in the 1920s when a second-hand one did not deliver enough power for a forge that he liked to use. It was on his father's ranch, so he took a fan off another windmill and the rear axle from a Model T to build another aerogenerator. It worked, but not very well.

Jacobs decided that when the wind was strong most of the blades retarded rather than accelerated the rotor. He reduced the number of blades then to three, and made them like airplane propellers but changed the pitch. With that rotor he got three times as much electrical energy as he could from a windmill with a multi-bladed fan the same diameter.

That helped, but there were many more problems. By figuring things out for himself, Jacobs wound up with more than two dozen patents. One was on a flyball governor that he placed at the hub of his rotor to feather the blades in high winds. He tried both solid metal and hollow aluminum blades, and settled for vertically grained Sitka spruce that he went to the West Coast to select personally. When the brake bands on his machine froze, Jacobs found that he could dispense with them and rely on a tail vane to keep the rotor from running wild when the wind was fierce, if he attached a spring to the vane.

By scouting around Jacobs was able to buy storage batteries and towers adequate for his engines, but he was unable to find a generator that operated satisfactorily when driven by his three-bladed propeller. To match the propeller, Jacobs designed and built a bigger generator, but it sparked too much. So he invented a new kind of brush for it, and a "reverse current relay" circuit. He was determined to build an aerogenerator that would last for many years in almost any environment, and continued to improve his engines for about ten years.

After making and marketing some machines in Montana, Marcellus Jacobs and his brother moved to Minneapolis to open a factory there. By 1937 they were producing wind-driven power plants in which they made no significant changes for the next twenty years.

Their 2,500-watt, 32-volt plant sold at the factory for about $490. A 50-foot tower and a bank of good glass-cell, lead-acid storage batteries for use with it cost about $540. The Jacobs brothers' factory employed up to 260 workmen and produced more than $50 million worth of aerogenerators before it closed.

On the western plains where the wind might blow from 7 to 20 m.p.h., a farmer sometimes could draw several kilowatt hours of energy for his home or shop with a Jacobs generator. They were equally helpful on many an island, in a mining camp, and at a weather station in the Arctic. Admiral Richard E. Byrd took one to Antarctica, and many years later his son found its three blades still revolving. A Jacobs machine shipped to Ethiopia in 1938 needed no parts replaced until 1968.

By the 1950s, however, the business was no longer profitable. It was easier and cheaper by then for American farmers to buy current from a central plant, even in thinly populated states. So Marcellus Jacobs retired to Fort Myers, Florida, where, like Tom Edison, he went right on inventing. To save the aquatic life in a coastal canal near his home, Jacobs has installed and patented gates that keep the water fresh by controlling the flow in and out of the canal.

As a lively, dapper senior citizen, Jacobs has repeatedly urged the federal government to help finance the development of a new small "wind energy package." But he considers the direct current machines he once produced "a thing of the past" today. If he were still in the windmill business, Jacobs has said, he would be concentrating his attention on larger units capable of feeding alternating current directly into the utility companies' existing lines.

Martin Jopp, one of Jacobs' competitors in Minnesota, has also stopped making home lighting plants. He now has a small shop in Princeton, Minnesota, where he rewinds motors, repairs pumps, and sells electrical appliances.

Jopp proudly recalls being asked to dismantle a windmill that he produced for a neighbor who used it for twenty-five years without ever having to climb the tower to repair the mechanism. Jopp still has four windmills on his farm, including a 3,000-watt, 110-volt, direct-drive machine on a 75-foot tower that keeps storage batteries charged for him. But, like the whole country, most engines sometimes sputter, miss, and slow down, and Jopp no longer bothers to keep all of his own windmills running.

## A CHARGER STILL SOLD

An American searching for a wind-driven generator to light his Christmas tree in the early 1970s could find only one make and size still manufactured and nationally advertised in the United States. It was a battery charger that the Albers brothers, John and Gerhard, originally designed and built to run a radio and light a couple of bulbs in their home on a farm near Cherokee, Iowa.

Their Wincharger was made in Sioux City, Iowa, an old town on the Missouri River, in a factory that men who grew up there in the first half of our century remember for having provided considerable employment. The Zenith Radio Corporation acquired the business after it was moved from Cherokee to Sioux City, and the Wincharger Corporation later became a Dyna Technology, Inc., division called Winco. It formerly made 6-, 12-, 32-, and 110-volt direct current chargers for storage batteries, but by 1975, it produced only the 12-volt model. This included a completely wired instrument panel and cost a little more than $500 when delivered to Washington, D.C., without batteries.

The Wincharger has a two-bladed wooden propeller 6 feet wide on the hub of a four-pole generator 7½ inches in diameter. It comes with a four-legged tower 10 feet high that you can perch on a roof or a platform. The whole machine weighs 134 pounds, and is kept facing the proper way by a big vane behind the generator (see page 197).

The Wincharger is a sturdy metal machine with a patented air brake operated by centrifugal force. Flaps on the tips of two arms between the propeller blades open automatically to retard the rotor when the wind exceeds 23 m.p.h. Marcellus Jacobs disdained this kind of governor, but it has been retained as a distinctive feature of the Wincharger because it reduces the effect of gusts on the generator.

At wind velocities between 3 and 23 m.p.h., a Wincharger's propeller revolves from 270 to 900 times a minute, and the generator delivers enough current to charge batteries. The company's specifications promise that from 20 to 30 usable kilowatt hours per month can be obtained from wind speeds averaging from 10 to 14 m.p.h.

The "rating" that a manufacturer gives to an aerogenerator is usually the maximum current that it will deliver. It will produce fewer or no kilowatts when the wind exceeds or falls below certain velocities. At many places you would need a wind-driven generator with a higher rating than any now on the market to be sure of getting enough kilowatt hours from it for an all-electric house. But on a boat at sea, or in a cabin or camp far from other sources of electrical energy, a small machine can be very valuable.

Government agencies have bought Winchargers for lookout towers deep in forests and for meteorological and other observatories in remote parts of the world. A radio ham who has one can stay on the air despite electrical blackouts caused by hurricanes, landslides, volcanic eruptions, or anything else that disrupts service from a utility company. Some owners of country homes recently have bought chargers to keep burglar alarms activated regardless of what happens to transmission lines.

One winter in the 1940s a severe storm stalled two cars and an oil truck on a lonely Nebraska highway. A man in one of the cars heard a Wincharger whirring, and followed the sound to a locked farmhouse. He broke in and turned on a light. Nine men, women, and children then plowed through the snow to the house from their stalled cars. One of the fellows milked a cow and fed a dozen hungry calves in the barnyard. The oil truck's driver stayed in his cab and lit flares to keep himself warm as long as he could. When the storm finally ended, the truck driver was found dead, but the people in the farmhouse all survived.

Hugo Gernsback, a science fiction writer who helped to popularize radio, pointed out in 1920 how hastily and expensively our country's power system was being developed. "What we are doing," he wrote then in *The Electrical Experimenter*, "is to mine the coal at enormous expense, move this coal to faraway points—the average distance being over 1000 miles for every car of coal . . . we block railway traffic with our coal trains. . . . Only one or two per cent of the energy confined in the original coal is converted either into power or into heat. . . . As manpower gets scarcer and costs rise, there will finally come a time when we must turn to other sources for our power."

Few Americans persisted in relying on small wind generators and banks of batteries for electrical power as long as Caesar Aimone at Iron Mountain, Michigan, and John Lorenzen on a farm not far from Des Moines, Iowa. In the 1970s newspapermen found that Aimone had six aerogenerators producing up to seven kilowatts for him, and Lorenzen was running one of his state's most highly mechanized farms without paying a utility company a nickel.

Lorenzen had drawn all the current that he needed from three aerogenerators and a bank of 175 storage batteries for many years. The windmills provided current to keep the batteries charged whenever the wind blew more than three miles an hour. Lorenzen greased the engines once a year and put a little rainwater in the batteries every few months. He had a workshop full of power tools, a large television set, and all the usual appliances, including an electric razor —and reported that the only time he had ever lacked enough electricity to get by was the night that lightning struck his electric fence.

In Europe the Danes especially appreciated aerogenerators because of their dependence on other countries for fuel. Professor P. La Cour's experiments with windpower at the Askov High School in the 1890s led to the formation of a Danish Wind Electricity Company in 1903. By 1914 Denmark had thousands of windmills that kept things going despite a wartime blockade of the country's ports. Small plants lightened the people's hardships again during the Second World War, and in 1942 a 200-kilowatt plant was built at Gedser to feed current into transmission lines.

Several more powerful plants than any constructed previously in the United States were built in other European countries, and in Germany Professor Hermann Honnef envisioned truly gigantic wind-driven generators in 1932. He proposed that five 6-bladed rotors 250 feet wide be mounted on each of several towers 1,000 feet high. But three successive German governments and the Marshall Plan's administrators rejected Honnef's ideas as too visionary.

Between the world wars, the Russians were the boldest of all the European experimenters with machines to produce electrical energy. At Yalta, where peat was still being burned as fuel, the Communists constructed a tower 100 feet high on a bluff overlooking the Black

Sea, and armed it with two 100-foot blades to drive a 100-kilowatt induction generator.

A motor governed by a vane on the tower kept that big turbine facing the wind properly, and the generator fed current into a 6,300-volt transmission line, working together with a 20,000-kilowatt steam plant twenty miles away at Sevastopol. The wind-driven generator's output was greatest when the wind blew 24.6 m.p.h., and the mean annual velocity there was only 15 m.p.h. Even so the plant was reported to have produced 48.4 kilowatts in the windiest month of the year and 18 kilowatts in the calmest months, and thus to have delivered a total of 279,000 kilowatt hours in the course of a year.

That was sufficient to excite many engineers, and was a direct challenge to the United States' technological leadership.

## XII

# Winds from War

What's here, beside foul weather?
KING LEAR, ACT THREE, SCENE 1

PALMER COSLETT PUTNAM conceived the most powerful electric plant that the wind has ever driven. It ran briefly on a knoll near Rutland, Vermont, to feed current into the whole New England network of transmission lines. Although its sails were visible for twenty-five miles, that great plant was a promptly forgotten victim of the winds from World War II.

Putnam was a 1923 graduate of the Massachusetts Institute of Technology who became a pioneer in energy science when a utility company's bills for lighting his Cape Cod home shocked him. He thought that the wind should provide all the current that he needed there, and maybe a few more kilowatts for him to send back to the power company for credit. Putnam's father was a famous New York publisher, and his mother was a noted author and scholar. He had studied at the Technische Hochschule in Munich and at Yale University as well as M.I.T., served with the Royal Flying Corps, traveled widely, and become an enthusiastic yachtsman. Few men were as well acquainted with social and business leaders, and eminent scientists and engineers in Boston and elsewhere.

The wind machines then producing a few watts for some farmers, of course, were much too small to satisfy Putnam. He reviewed the progress made in building windmills. The big sails of Dutch windmills had rarely revolved faster than the wind was blowing. The tips of the blades on American windmills went twice as fast, and with propeller types of blades builders had attained working ratios of five or more. A high number of revolutions per minute was helpful in generating electricity, but no one seemed to have comprehensively determined the most economical dimensions for a very large wind turbine.

Putnam considered the Russians' approach to the problem bold and practical; they had begun, however, with an induction generator rather than one that could be synchronized with other power plants, and their machine at Yalta was crudely constructed. Roofing metal was used in the Russians' rotor, and the main gears were wooden. Surely, Putnam thought, with the greater resources in the United States, a better wind turbine could be built here.

Elisha Fales, one of the first experimenters with airfoils on windmills in this country, lent Putnam a small two-bladed rotor that he mounted on a 60-foot tower and used as a kind of gauge to study the wind. Another friend interested him briefly in a Savonius rotor that was being considered for use on windmills on Colonel Henry Huddleston Rogers' estate on Long Island; but Putnam concluded that long vertical airfoils would be better for his purpose.

The surface exposed must be larger for a wind turbine than for a water turbine because air is thinner than water. Hydroelectric and thermal plants were generating alternating current for the New England utility companies, and Putnam proposed to produce current in phase with them. This meant that both the rotor and a synchronous generator would have to be especially designed for a very large wind turbine on a windy site. So, too, would a structure to support it.

Putnam consulted many of our country's leading authorities on aerodynamics, electrical generators, structures, meteorology, ecology, and construction costs. With their help he succeeded in launching an unprecedented historic project. Although nothing is left of the great machine that resulted from his efforts, his diagnostic obituary of it is still a timely textbook on an aerogenerator builder's many problems.

The Smith-Putnam turbine weighed 250 tons and was erected on a 2,000-foot hill called Grandpa's Knob in the Green Mountains west of Rutland, Vermont. The tower was 110 feet tall, and the operators of the plant rode to and from the control room on an elevator. Up there the wind spun two long bright metal airfoils in a rotor 175 feet in diameter. A specially designed generator came on line when the wind blew 17 m.p.h., and produced 1,250 kilowatts when it reached 30 m.p.h. This was enough for a small town, and machines like this could help lighten the growing loads on the utilities.

A complete roster of the participants in the Smith-Putnam project would be virtually a Who's Who of the most able scientists and engineers in the 1930s. The dean of engineering who became President Roosevelt's wartime research director, Vannevar Bush, put Putnam in touch with academic celebrities and the movers and shakers in American industry. The project, as Dr. Bush wrote later, demonstrated "the ability of complex science and technology to focus a score or more of specialized skills on the various aspects of a problem." This was done, moreover, without any subsidies from federal or other government agencies.

### GETTING THE GIANT TOGETHER

The times had seemed propitious. The scars from the Great Depression were healing; economists were debating new theories, and investors were regaining confidence. Putnam got estimates from steel makers and other manufacturers of the probable cost of producing the big aerogenerators he envisioned in lots of 100, and the decision to build an experimental model was made in Boston in 1939. That was the year a world's fair opened in New York, and also the year that a second world war began in Poland.

Thomas F. Knight, a coastal sailor and vice president of General Electric, had become one of Putnam's most helpful allies. General Electric agreed to develop and provide a synchronous generator that could be used in conjunction with New England's hydroelectric and thermal plants. Knight also helped to interest other wealthy and influential companies in the venture. He knew there were few good sites left in New England for hydroelectric plants, that builders of turbines were looking for new markets, and that utility companies were worried about the growing demands on reservoirs from which they were drawing water to generate current.

A leading builder of hydraulic turbines, the S. Morgan Smith Company in Pennsylvania, was a family owned enterprise with an exceptionally able staff of engineers. Walter Wyman, the New England

*Blade of the Smith-Putnam wind turbine being moved to its hilltop site.*

Public Service Company's president, lived on a windy ridge and was receptive to the idea of developing a new source of power. After considerable study, the Smith company agreed to build a big wind turbine, and Wyman arranged for a Vermont utility company to feed the current that it would generate into transmission lines serving several states.

Professor John B. "Bud" Wilbur, who later headed M.I.T.'s Department of Civil and Sanitary Engineering, became chief engineer for the project. Dr. Theodore von Karman of the California Institute of Technology, and other noted aerodynamicists, helped the designers solve many difficult problems. The famous Norwegian meteorologist, Sverre Petterssen, and fellow scientists at Harvard, M.I.T., and elsewhere, began a thorough study of the wind in the Green Mountains.

The war in Europe distracted and rushed nearly every participant. The S. Morgan Smith Company was soon swamped with war work, but succeeded in assigning completion of the aerogenerator to other experienced builders—including the Budd Company in Philadelphia, which built stainless steel railroad cars; the Wellman Company in Cleveland, which produced heavy materials-handling equipment; and the American Bridge Company, which had erected many huge structures.

The war in Europe went badly; it seemed increasingly certain in 1940 that the United States would be involved. If big forgings for the wind machine were not ordered immediately, it might be impossible to get them made. So they were ordered on the basis of approximate and fairly rough estimates of stresses, before the big turbine had been fully designed. Grandpa's Knob was likewise chosen as a convenient site to get on with the project, even though the meteorologists did not yet have as much data as they wanted.

The schedule called for cumbersome heavy parts to be hauled up the side of the mountain in bitterly cold weather when the curving road was frozen solid. The main girder with the driving shaft and bearings weighed so many tons when placed on a trailer that two Caterpillar tractors were needed to tug it. At a hairpin turn the girder fell off the trailer and landed upside down under a snowbank in a rocky crevice. Workmen had to struggle for three weeks to get that girder back on the road.

Each one of the two blades for the turbine was 11 feet wide, nearly 70 feet long, and weighed 8 tons. Those blades, the generator, and other heavy parts had to be carefully balanced—like little millhouses on single posts centuries earlier—on top of the tower. After the first blade was attached to the high hub, it had to be held in a fixed position until the second one was attached. But all such challenges to men's strength and skill were met, almost cheerfully at times.

Palmer Putnam was called to Washington to help Dr. Bush in the Office of Scientific Research and Development before the plant he had proposed was finished. Engineers assigned to test the machine opened a field office in Rutland and escorted many distinguished visitors up to the control room with mechanics and inspectors. A General Electric operator began letting the machine run slowly, and more weeks of testing and tinkering were necessary before it was allowed to run at full speed. Then on Sunday night, October 19, 1941, it fed energy from a gusty 24-mile-per-hour wind into the utility companies' lines, for the first time that this ever was done in the United States.

"If this wind-power project proves as successful as engineers working on it believe it will," *The Boston Post* had reported that spring, "the vision of thousands of similar stream-lined scientific windmills operating on the wind-swept summits of New England's mountains and high hills is not too fantastic to contemplate. . . . A few well-aimed aerial bombs striking New England's largest power plants would cut off the supply of electricity to a widespread area of this in-

*Blade being fixed in place for rotor on top of the tower.*

dustrial section. But wind turbines, distributed through the hills and camouflaged to blend with the scenery would be much less vulnerable to destructive attack from the air than equivalent generating capacity concentrated in a single large plant."

## HOW A BLADE BROKE

Testing continued after the Japanese attacked Pearl Harbor on December 7, 1941, but fewer people paid attention to what was happening on Grandpa's Knob. When fully adjusted, the turbine ran smoothly and frequently delivered 1,000 kilowatts. This was a test unit that purposely had not been placed at the point of maximum output in New England lest high continental winds make constructing and testing it too difficult. Yet it once drew 1,500 kilowatts from a 70-m.p.h. wind, and when idle it withstood gales up to 115 m.p.h.

A bearing failure halted use of the machine in the winter of 1943. By then the Allies were demanding that their foes surrender unconditionally, and the war delayed replacement of the bearing. Design analyses had shown that the blade shanks over which the spars of the blades should fit were too small, and the connections permitted greater stress concentrations and more fatigue than the engineers considered acceptable. No one could be sure that the connections would fail, but neither could anyone promise that they would not.

Should the whole project be halted until after the war, when new and larger forgings could be obtained? Or should the testing be resumed? The S. Morgan Smith Company decided to go ahead, with electric strain gauges mounted at certain critical areas.

Another omen of trouble was discovered during the long shutdown caused by the failure of the bearing. The blades were feathered and held vertically then, and hairline cracks appeared at some points on their skins. These apparently resulted from the vibration and shaking of the blades while they were locked in one position. One suggestion was that those cracks be prevented from growing by welding bulkheads near the shank-spar connections, and this was done.

By then everyone realized, Chief Engineer Bud Wilbur recalls, that "we were living on borrowed time." He recommended that the plant only be run long enough to complete certain tests before discontinuing operations. It could be strengthened more after the war ended. Beauchamp Smith, president of the company responsible,

*The turbine completed on Grandpa's Knob.*

agreed, and the tests were resumed.

The wind was blowing steadily about 20 m.p.h. on a moonlit night in March 1945 when one of the two 70-foot blades left the turbine and landed down the side of the mountain. At six o'clock that morning Bud Wilbur telephoned the man who had conceived and promoted the construction of the great turbine to report: "Put, we've had an accident. We've lost a blade, but no one is hurt, and the structure is still standing."

Only one man was aloft in the control room when it happened. The jolt knocked him down, and he fell again when he tried to leap over the 24-inch rotating shaft to stop the mighty machine. But he regained his feet and overrode automatic controls that already were functioning to feather the remaining blade and bring the machine to a full stop in ten seconds.

The Smith-Putnam windmill never ran again, but the wind survey begun before it was built was continued. The meteorologists had placed instruments at many Green Mountain sites, obtained data from balloons, and festooned a 185-foot skeleton steel tower at Grandpa's Knob with so many gauges that people called it a Christmas tree. Anemometers heated by gas and later by electricity were developed for use on it. Ecologists measured the height and deformation of trees and noted the icing on them. The terrain also was modeled for aerodynamicists to study the effects of air rushing over it in a wind tunnel. This was the most thorough study of the wind in the region that ever had been made, and much was learned from it.

Participants' memories and published articles have differed about the troubles with the machine. In Dr. Von Karman's autobiography he likened his experience as a consultant for the project to what happened when a finicky man ordered a steak in a restaurant. "Cut the steak exactly two and a half inches thick," he told the waiter, "add a pinch of pepper and a half teaspoon of salt. Keep vegetables on the side of the plate. Get it?" The impatient waiter nodded, turned around and yelled to the chef: "Steak and french fries for one."

The engineers who built the big turbine had foreseen trouble, and hoped that "modified and cleaned-up" units would be built on other sites in the Green Mountains after the war. This would have required more money, however. The S. Morgan Smith Company had spent about $1.25 million on the project, and its funds were limited. Estimates of the cost of building more wind turbines indicated that large blocks of power could be generated much more economically in other ways.

Nuclear energy was a factor in the decision to dismantle the turbine on Grandpa's Knob. The bombs that ended the war in Japan suggested that nuclear power could also be used for peaceful purposes. Its potentialities excited both engineers and investors. So Beauchamp and Burwell Smith persuaded Putnam merely to review what had been done in a technical treatise that might be helpful some time where winds were strong and fuel was scarce.

In an introduction to Putnam's report, Vannevar Bush wrote: "The great wind turbine on a Vermont mountain proved that man could build a practical machine which would synchronously generate electricity in large quantities by means of wind power. It proved also . . . that at some future time homes may be illuminated and factories may be powered by this new means."

### MORE PLANS THAN ACTION

Putnam recommended that a national survey of the wind's energy be made, and that the government support efforts to use windpower. The War Production Board thought windpower might be most helpful at military bases overseas where there were few other resources, and sponsored studies of aerogenerators for those bases.

New York University engineers then proposed that a 100-kilowatt wind machine be built with more standard components than had been used in the Smith-Putnam giant. Its two blades would be only 2 feet wide and 24 feet long, made of Sitka spruce and held together by an aluminum alloy. That rotor would spin a hollow steel shaft 6 inches in diameter to drive a three-phase 220-volt, 60-cycle induction generator.

A diesel engine would be used to excite and start the generator and pick up the load when the wind died down. The tower would be only 72 feet high, and the machinery could be enclosed in a wooden or sheet steel housing. By building such plants overseas, the designers thought, the government could save fuel oil and reduce shipping costs. But there still seemed to be plenty of oil underground in this country, and the government did not build any aerogenerators.

Percy H. Thomas, another M.I.T. graduate, succeeded Palmer Putnam as this country's leading proponent of using big turbines to generate electrical current from the wind. Thomas wrote a series of scholarly monographs for the Federal Power Commission that its spokesmen were still quoting in the 1970s. In the first one he suggested that a machine be built to generate five times as many kilowatts as the one on Grandpa's Knob. In a second report to the commission Thomas continued to favor very large machines, but in a third one he modified his proposals.

Those studies resulted in more hemming and hawing than action in the 1950s. William E. Warne, Assistant Secretary of the Interior, and others who testified at Congressional hearings in 1951, urged that the government construct a 7,500-kilowatt pilot plant. But too many legislators remembered the antics of the little windmills used in the midwestern states, and were loathe to see the taxpayers' money spent in that way.

The Federal Power Commission did not publish Thomas' fourth and final monograph on windpower until 1954. Thomas, who had retired by then, proposed a new plan in this report. He conceded that it would be impractical to build a big enough wind machine to deliver, say, 100,000 kilowatts to a network of high-voltage transmission lines. But he pointed out how greatly the wind's speed often varies within a small area. By combining the output from many generators at different sites, he argued, electrical energy might be generated continuously and economically enough to interest the utility companies and reduce their use of fossil fuels.

Putnam quoted a noted scholar, J. B. S. Haldane, in the final chapter of his book. When writing about the future, Haldane had once said "it was characteristic of the dwellers on the earth that they never looked ahead more than a million years and the amount of energy available was ridiculously squandered."

In the quarter century after atomic bombs first exploded, few Americans supposed we would ever have to conserve energy. But we must now, and engineers have revived Thomas's idea of building fleets of aerogenerators to delay exhaustion of other sources of energy.

# The Windmill's Defenders

Woodman, spare that tree!
   Touch not a single bough!
In youth it sheltered me,
   And I'll protect it now.

WILLIAM POPE MORRIS IN NEW YORK MIRROR IN 1817

I N  T H E 1870s many Americans objected to telephones because they felt foolish talking to a faceless gadget. In England in the
same decade, William Morris, a socialist poet, founded a Society for
the Preservation of Ancient Buildings. Many more organizations are
now resisting changes in familiar environments. In Washington, for
instance, we have a Don't Tear It Down society, and Kansas has a
group battling to Save the Tallgrass Prairie.

Windmills were so plentiful until this century however, that few
people united to preserve them, in either Europe or America. What
happened to an old windmill was strictly its owner's business. A
family sometimes saved one for sentimental reasons and towers often
were left standing because they could be used as homes, shops, or
sheds. On the Great Plains farmers with electric pumps sometimes
left a windmill's skeleton in a field to show prospective buyers that
the land had long been productive.

Even in Europe intensive campaigns to protect old windmills did
not gain momentum until the 1920s. Dr. F. G. van Tienhoven then
published a pamphlet urging the Dutch to save their windmills because of their historic importance, and *De Hollandsche Molen* was
organized. Its founders hoped to prolong the usefulness of windmills
by improving the mechanisms within the old towers, like surgeons
replacing parts of a human body. Unfortunately rejuvenation projects
like this cost so much that few owners could afford them. But tourists'
exclamations of delight when they saw a fine old Dutch windmill

still toiling in its natural setting awakened business-minded men. Merchants then persuaded national and local officials to take steps to prevent windmills from being torn down or moved to make way for new buildings.

Windmills became national monuments in the Netherlands, and a Royal Decree in 1945 required a government commission's consent to demolish or alter one. Since 1961 all windmills in the Netherlands have been officially registered, and since 1970 a government department has been responsible for restoring them when possible. *De Hollandsche Molen* became the world's most successful windmill preservation society; in 1973 the Queen of the Netherlands, and celebrities from all over the world, helped observe its golden jubilee. Municipalities own more than two hundred of the country's windmills today, and foundations manage and operate a hundred more. Many more survivors of the nine thousand windmills that the Netherlands had in the nineteenth century have been saved, although only about a tenth of the survivors have full-time millers.

Convinced that "rest rusts," *De Hollandsche Molen* has sponsored a Guild of Volunteer Millers to teach "serious, dedicated mill-lovers" how to tend a windmill in their spare time. Completing the course takes most people a year or more, and a diploma is not a key to a mill to live in or loaf around on weekends. Tom Rouwenhorst, a director of one of Holland's largest shipping companies, took the course and found that "there's so much to learn that there's no time for romanticism." The millers' guild now has about 24 instructors and 500 student members, of whom some 120 have passed a rigorous final examination.

Eight distinct kinds of Dutch windmill—ranging from tiny meadow drainage machines to gristmills with rotors 96 feet wide—have been kept operable. To display them, the National Tourist Office offers visitors six different windmill tours. On one tour they may see sixteen windmills running simultaneously on Saturday afternoons.

Neighboring countries have turned to their artists for help in attracting tourists similarly. In 1962 the Belgian *commissaire general à tourisme* sponsored publication of *Les Moulins à Vent de Belgique* by Robert Desart, a beautifully illustrated guidebook to windmills that go back to 1390.

Herman A. Webster, an American artist, was especially helpful in the establishment in France of a *Société des Amis des Vieux Moulins* in 1928. He was a Chicago newspaper publisher's son whose eyesight was impaired by a gas attack in World War I. When he recovered, Webster stayed in Europe to depict the Old World's splendid structures. Six of his sketches and drawings of European windmills were hung in our country's National Collection of Fine Arts galleries in Washington in the 1970s.

In northwestern Germany less than fifty of the nine hundred windmills in use there half a century ago are maintained, according to a Schleswig-Holstein society striving to preserve old wind- and water mills. There, as in the United States, many have been saved by converting them into homes.

Few people have ever gone to as much trouble to have a luxurious windmill residence as Helmut Schmook, a wealthy German plastics manufacturer. When Schmook acquired his windmill every beam and crevice in it was so covered with fine white flour that he needed a canopy over his bed to keep himself from looking like a ghost in the morning. Schmook spent a year of weekends getting rid of that dust and the flour worms before he got the old machinery running smoothly again. Finally he and his guests were able to relax and enjoy seeing the mill's big sails whirl past the window without having to wash up afterwards.

Thanks to the determined efforts of many individuals in many countries, architects and engineers have discovered that windmills are well worth studying as examples of past technological change. Hence many men have joined artists, patriots, and other nostalgic citizens in efforts to preserve windmills for professional reasons.

## WORLDWIDE CO-OPERATION

The International Molinological Society (TIMS) now coordinates the activities of groups and individuals interested in milling's origins in many countries. "Molinism" was a doctrine preached in the sixteenth century which held that a man's free co-operation with others enables him "to perform a good act with God's grace"; and the cross that supported a runner above a bedstone in a mill was called a "moline."

The scholarly founders of TIMS chose a symbol as well as a name with a long history. It is a drawing of a post mill copied from one on a British halfpenny. That coin was never authorized, but it was widely used in the seventeenth century when the government failed to provide enough halfpennies for small businesses to survive. J. Kenneth Major, an architect for the City of Westminster, has served as secretary of TIMS since it was organized in the 1960s in Denmark.

The Danes have long been noted for their use of the wind. The world's first wind tunnel was built in their little country in the 1890s to study wind-driven wheels. Those wheels ran generators that produced record numbers of kilowatts of direct current for the Danes in the first half of this century. Direct current was used extensively on their peninsula and islands before alternating current was brought by cable from hydraulic plants in Sweden. When TIMS was organized the Danes remembered well how helpful and beautiful their old windmills had been when Denmark's ports were blockaded and they could not import fuel for thermal power plants.

The members of TIMS are interested in every aspect of the origin, use, and fate of old water- and windmills everywhere. The founders agreed therefore to meet for a symposium every fourth year in a different country to promote and disseminate knowledge of the milling industry's development and contributions to technological progress. Published proceedings of two of those meetings have been fascinating additions to technical literature, clearly written, carefully edited, and finely illustrated and printed. Technical writers and editors in the United States should be more familiar with them.

The Portuguese Society of Friends of Windmills was host to TIMS' second meeting, and the third one was postponed from 1972 to 1973 so that the members could participate in the celebration of the golden jubilee of *De Hollandsche Molen* in The Netherlands. Wooden mills on stone bottoms can be kept running indefinitely if wooden parts and sails are replaced when weakened, and in both Portugal and Holland the molinologists found a variety of machines dating back to the 1600s which were well kept, spotlessly clean, and still used.

In Lisbon the city owns two windmills on a hilltop surrounded by stark new apartment houses. The attendant miller will permit you to enter one for a few cents, but you may have to keep your head down to avoid being conked by a heavy sail. Stone steps lead up to a turret room that resembles the interior of a great ship. Sails sweeping by a tiny window make the light flicker madly inside the tower, beams creak and squeak, and corn is jogged down into the millstones to be whirled out as meal.

TIMS hopes to locate, date, and register historic windmills in every country. One enthusiastic member, J. Walton, has mapped the sites of mills run by horses, water and the wind on the southern tip of Africa in the 1650s. Another, Anders Jesperson, has compiled an index to Denmark's mills that has been published under the imprint of Bibliotheca Molinologica. The older a windmill is, the more likely it is to fascinate a TIMS researcher.

You need not go to other countries, however, to enjoy the same pastime. An amateur molinologist can study the world's past and present windmills in an armchair with a magnifying glass. More than forty countries have depicted their favorite windmills on their postage stamps. Both the northern and southern hemispheres are well represented.

Russia, China, the United States and all of the European powers have put out windmill stamps. So have little countries like Monaco, San Marino, and Liechtenstein, and islands throughout the seven seas, including the Takelaus in the South Pacific and the Maldives off the tip of India.

Don Quixote and early European windmills still appear on many stamps. One from Surinam shows a Dutch windmill's four big sails at Nieu Amsterdam on the Guiana Coast of South America. American-style windmills are well portrayed, too; a stamp from Somalia shows one at a desert water hole where a caravan has paused. Robert Daly, a Toronto collector, has more than ninety different windmill stamps and has heard from several dozen other searchers for them. Lee Saegessor, a historian in NASA's Washington office, has found that collecting such stamps is a pleasant way to review the origins of technological advances and how they have spread.

### REMINDERS EVERYWHERE

People remember windmills for many reasons. Charles A. Mosher grew up beside a windmill factory and sang about a windmill in a high school operetta. "Where there's a wind there's a way," he

145 ॐ

thought later when he was confronted by an energy crisis as a Congressman from Ohio on the Science and Astronautics Committee.

To list all the songs, children's books, theatrical performances, and advertisements in which windmills have been pictured would be a job for an electronic computer. Collecting jewelry, dishes, clothing, seals, labels, etc., showing windmills is like collecting sea shells—more can always be found. The decorative quality of windmills accounts, no doubt, for the multitude of tiny windmills in American lawns and gardens. Home handymen often make them, and at least two firms now produce windmills from 4 to 8 feet high to sell by mail to others who yearn for one.

Windmill watchers may find the wind's power being used in surprising new ways. In Colorado biologists interested in preserving wildlife recently have put up big windmills to keep fish alive in frozen lakes by oxygenating the water under the ice. And Sears' 1974 Suburban Farm and Ranch Catalog pictured a Pondmaster manufactured in Joplin, Missouri, to keep small ponds from freezing by stirring the water. Those machines have three "sails" made of metal bent into half circles and mounted on a vertical axle. They begin to turn whenever the wind reaches 3 m.p.h. and can withstand gusts up to 60 m.p.h.

An even more impressive sight greets winter visitors to a wildlife preserve in Nebraska's sandhills. There they may come across a dozen or more tiny wind wheels busily whirling on Crescent Lake. Ron McCartney, an avid ice fisherman who lives on a nearby ranch, sets those little windmills out to jiggle the bait on his lines under the holes he cuts in the ice. McCartney uses fans from the heaters of junked automobiles as rotors and baits his hooks with strips of liver. The midget windmill on the "tip-up" he places over a hole in the ice makes the liver look like live bait to a fish. When it bites the hook, an automatic reel on the tip-up lets the fish run with the bait until it tires, then retracts the line. McCartney insists that he has caught more and bigger fish than most ice fishermen since he began using the wind to wiggle the bait.

A TIMS survey of the milling industry in the United States probably would not include contrivances to save or catch fish, but it would show historic windmills that have been restored and reconstructed for parks in many of our states. Harley and Barbara Stroven started one of the first of these parks at Fremont, Michigan, in the 1930s. There you can now see the kind of multi-bladed fanwheel that Daniel Halladay, the Connecticut inventor, put on the windmill he patented a century ago, and learn more about that and other windmills in their Halladay Mill House Museum.

One of the newest windmill parks, along the Interstate Highway across Nebraska, was started by a Chamber of Commerce to increase the attractiveness of a trailer campsite, near Gibbon. There, the Indians waded across the Platte River until the Union Pacific's little locomotives, with billowing smokestacks and enormous cowcatchers, stopped at the place for the water pumped by a big windmill.

When designated as a recreation area and campsite for tourists a few years ago, this tract of land was full of weeds and sandburs. No trace of a windmill was left, although the site was still called Windmill Park. So when the newly planted grasses, shrubbery and shade trees induced travelers to pull off the road, many of them embarrassed people in Gibbon by asking them what had become of the windmill.

Mr. and Mrs. Louis Drake of Arcadia ended that by giving their friends in Gibbon an old wooden mill to move to the new park. Winston "Windy" Lindsay restored and repainted it to justify the park's name until a railroad windmill could be erected. To finance that, D. E. "Mac" McGregor and other local businessmen organized a WPA (Windmill Preservation Association) and attracted dues-paying members by offering them certificates stating that they were POWs (preservers of windmills).

Windy found the remnants of a Burlington railroad windmill in Colorado, where it was called The Big W. It was a mess, but he recognized it as an Eclipse machine suitable for the site, got an old parts catalog, and tracked down duplicates of missing parts. A local contractor, Dean Brown, loaned Windy a boom truck to help him construct a tower high enough for The Big W's 20-foot-wide wind wheel.

By then the WPA had so many members throughout the country, and so many motorists were pausing to see the park, that the Gibbon Chamber of Commerce began to think of more ways to improve the area and make it a national windmill museum. In it they hope eventually to have restorations and reconstructions of examples of several different kinds of windmills that helped to make the Great Plains habitable and beautiful.

ALLIES IN AMERICA

The United States does not have any national organization such as *De Hollandsche Molen*, but it does have a growing Society for Industrial Archeology (SIA). The society was organized in 1971 at the suggestion of the director of the old Slater Mill Museum in Rhode Island. The Smithsonian Institution's curator of mechanical and civil engineering, Robert M. Vogel, edits a newsletter that alerts members to threats to factories, depots, bridges, mills, and other industrial structures. Vance Packard, a popular writer's son, serves as secretary of the SIA at the William Penn Museum in Harrisburg, Pennsylvania.

Another society especially concerned with mills of all kinds was chartered in Maine in 1972 as the Society for the Preservation of Old Mills. "All my life," the society's president explained, "I have enjoyed my mill hobby, and I knew there were many like me. I have checked out mills, sites, dams, flumes, machinery and artifacts, looked through museums, watched millers at work and talked with them, looked up histories, and thought about these things; and we often talk and swap information and become friends." The Society for the Preservation of Old Mills publishes a quarterly, *Old Mill News*, which is exactly what its name implies, and provides an excellent source of information about old windmills.

The most ambitious of the many local projects to preserve windmills in the 1970s was reported from San Francisco, where Mrs. Eleanor Rossi Crabtree, the daughter of a former mayor, spearheaded a drive to restore two gigantic Dutch mills in the city's Golden Gate Park. When erected in the early 1900s the mills were said to be the most powerful Dutch mills ever built, but they were used for only a few years. Both lost their sails while idle, and the towers were so neglected that it became dangerous to enter them.

Before the giant windmills were constructed, a newspaper described the Golden Gate Park as "a dreary waste of shifting sandhills where a blade of grass cannot be raised without four posts to keep it from blowing away." Sprinkling wagons hauled water to keep the dust down on the roads in the park because most people doubted that wells could be sunk so close to the sea. A Scottish gardener, John

*Giant Dutch windmill constructed in Golden Gate Park, about 1905.*

149 🦢

McLaren, became San Francisco's Park Commissioner in 1887, and finally persuaded the city fathers to sink a well and put a Dutch windmill over it in 1902.

That 75-foot-tall windmill cost $25,000 and raised 30,000 gallons of water an hour to a hilltop reservoir. Samuel G. Murphy, a wealthy banker, was so impressed that he contributed $20,000 for the construction of a second windmill five feet taller than the first. Other citizens donated cement, granite, and lumber to build what was hailed as the most powerful Dutch windmill in the world. With those machines providing fresh water, John McLaren turned the Golden Gate land into one of the most beautiful municipal parks in the United States.

The Murphy windmills' broad sails were spread on heavy wooden frames attached to masts 114 feet long, 2 feet square at the hub, and 8 inches wide at the tip. The prevailing 15-m.p.h. westerly wind spun those massive sails to pump 40,000 gallons of water an hour up to a lovely fresh water lake. High cypress trees surround that lake today on what was once a bare hilltop.

But a few years after those windmills went to work the pumps were electrified. The windmills' arms became eyesores that had to be removed. When Mrs. Crabtree was a child her father had often driven his carriage through the park so that she could see the sails rising and falling, and to her it was "a heavenly sight" that she has vowed to see again.

Mrs. Crabtree had raised $140,000 from garden clubs, historical societies, artists, and businessmen by 1974 for the John McLaren Society, organized to restore the two windmills. She became known in San Francisco's City Hall as "the windmill lady." Inflation was increasing the cost of realizing her dream, and the energy shortage was awakening the city's newspaper editors. *The Examiner-Chronicle* lamented the neglect of the powerful windmills and declared the time was ripe to put them into action again. The Seabees on Treasure Island volunteered to help, and the San Francisco Labor Council did not object. Engineers throughout the Bay area became interested, but tight budgets kept officials' feet dragging.

If completed, the restoration of those landmarks will surely inspire more communities to preserve their famous old windmills.

PART THREE

# Tomorrow's Windmills

## XIV

# Improving the Wheel

O brave new world
That has such people in't!
THE TEMPEST, ACT FIVE, SCENE 1

THE WIND WHIRLED new kinds of vertical wheels in the United States in the early 1970s that revived hopes of some day having a windmill on the roof generate enough electricity for a small home. The new sails weighed less than older rotors, needed less support, and ran very fast. Storing electricity might still be expensive, but the new rotors to produce it might cost less and be easy to put up. Monuments to committees are rare in American parks, but some of the teams that built these novel windmills soon received orders for them from park developers. Eager do-it-yourselfers also wanted plans and parts from which they might build their own aerogenerators.

One of the prettiest of the new machines was on a Princeton University campus, where windmills had previously been confined to lecture halls, coffee shops, and bars. Nothing like it had ever come out of that old school's many nooks and crannies before. Dignified professors insisted on calling it an "experimental advanced technology wind-driven turbine" rather than a windmill.

It looked like a pinwheel on a candy stick, intended to amuse children, when it was put up in an open field near the big Forrestal Laboratory. The tower was just an aluminum pole with blue and white stripes on it. The sails were flexible, like those on a sailboat,

*The Sailwing puzzled many visitors to Princeton's Forrestal campus.*

with orange and white stripes on them. The rotor was only 10 feet wide, and a simple tail boom and fin kept the sails flying around gaily in the wind's path. Even its maker had not thought of putting flexible sails on a stationary power plant until he saw an airplane land safely with them.

Thomas E. Sweeney, director of Princeton's Advanced Flight Projects Laboratory, had watched windmills out west when he was a boy. As a senior Princeton researcher, he had a 27-foot sailboat that he called the *Sailwing* for which he designed a new sail. It attracted limited attention until the Fairchild Corporation agreed to support more study of the sail for possible use on an airplane. Sweeney tested the sail in a Langley Research Center wind tunnel and elsewhere. It performed so well that it was tried on several types of small planes. While watching one of them land one day, Sweeney turned to a friend and asked, "Why not use this technique to build a better windmill?"

So they did. Its sails were sheets of "Dacron" polyester fiber. The straight leading edge and both the foot and tip of each sail were rigidly attached to a mast. A flexible cable held the cloth taut so that it was flat when there was no pressure on it from the wind, but when the wind was strong it curved the sheet into an airfoil.

No ribs were needed to keep this sail flat or curved, and the wind whirled the rotor faster than a heavier wheel. Yet the sails withstood gales, freezing rain, and heavy snowstorms for a year without being damaged. While observing that prototype's performance the Princeton experimenters also tested a sailwing mounted to be rotated horizontally rather than vertically. They then designed a larger machine, slightly different from the first one, and erected it in 1972.

This one's rotor was 25 feet in diameter, with two sailwings attached to a horizontal axle, and these sails were kept facing the way that the wind was going rather than in the direction from which it came. The rotor then ran a generator behind it that produced about 7 kilowatts from a 20 m.p.h. wind.

"We had just been through a series of very successful experiments with a high-performance sailwing on a light airplane," Sweeney explained, "so in a sense we had already done our homework. Excessive loading has always been a windmill problem because of the very high drag forces at high wind speeds. The energy in a 20-m.p.h. wind is eight times that in a 10-m.p.h. wind. Windmills have always had a tendency to fall down or become damaged in high winds, so we expected that an adaptation of airplane structural design technology would minimize this problem. It was clear at the outset that the sail-wing would also minimize the cost, and economy was another of our goals."

Although fuel prices had not greatly alarmed many Americans yet, backyard builders and searchers for an alternative energy source deluged Princeton with questions. "Defining the efficiency of a windmill is frustrating," Sweeney warned, "because so many factors must be considered." But an engineering grapevine buzzed with talk about a new inexpensive way to draw power from the wind wherever it was strong and reliable.

Other experimenters hoisted their own versions of sailwings to test them. In India a rotor was made by putting bamboo poles on a bullock cart wheel. They served as leading edges for flexible fabric sails, and nylon cords between the poles served as the trailing edges. That contrivance was intended primarily for irrigating small fields and supplying water for cattle and domestic use.

In the United States the demand was greater for electrical energy. Near the Grumman Aerospace Corporation's plant on Long Island the average annual velocity of the wind had been reported as high as 12 m.p.h., and a modest home often required less than 12,500 kilowatt hours a year. The company's Energy Systems Group became interested in Sweeney's machine.

In 1974 Grumman spokesmen reported that the company had been licensed to manufacture a Sailwing Wind Generator on an exclusive basis and had begun work on an experimental 6-to-12 kilowatt model. Flanagan's Plans, a firm in Astoria, Long Island, announced it would prepare and sell plans for building that machine to home handymen. But a year later the manager of Grumman's solar products division reported the project to manufacture and sell windmills was still "only in the discussion stage."

## WISCONSIN WINDWORKERS

The same November 1972 issue of *Popular Science Monthly* that called home handymen's attention to the Princeton windmill told them how to build another novel kind of rotor whirling in Wisconsin. It had three solid blades cut out of a block of expandable paper honeycomb with a hexagonal surface. This material cost only a few dollars, and airfoils could be made by expanding the paper, covering it with fiberglass cloth, then sanding and filling it.

Wisconsin was where a missionary had developed the Eclipse farm windmills in the 1860s, but by the 1960s most of that progressive state's towers for multi-bladed fans were dismal blots on the landscape. The wind had beheaded some, and vines had mercifully covered the rusting skeletons of others. Searchers found only a few that still ran; like one pumping water into a reservoir on Lewis Gade's place near Reedsville.

People gazed wistfully at such reminders of earlier times when students at the state university rioted in the 1960s; and some young people on a communal farm near Mukwonago began to build windmills again. Hans Meyer, an engineer, headed this research-oriented collective, and since 1970 its members have concerned themselves with "the development of structural systems that allow man to live in greater harmony with his world."

Meyer's followers included men like Tom Vonier, an architectural student who would later design museum exhibits for the American Association for the Advancement of Science. They called themselves Windworkers, and strove to promote "the transfer of information that allows the greatest learning and experiencing on the part of those giving and receiving." Artists among these technological missionaries wanted to work with new materials the way many modern sculptors do, so the choice of the paper honeycomb for a wind wheel's blades was not surprising.

Vonier recalls that the first idea considered was to intensify the current of air striking a wheel. In 1921 *Science and Invention* had pictured a power plant with a big horn on one side of the building to catch the wind and turn a turbine before the air could escape through another horn on the opposite side. That idea had been patented, and its potentialities intrigued other imaginative engineers. But the Windworkers concluded that a small windmill constructed that way would be difficult and costly for an amateur to build. A plain three-bladed vertical wheel, such as Marcellus Jacobs used on his famous aerogenerators, could be put up more easily.

The machine that Meyer recommended for do-it-yourselfers in 1972 was a handsome structure with a 10-foot rotor on a wooden

*Wind generator designed for home handymen by Wisconsin Windworkers.*

tower 12 feet high. The wheel faced the way that the wind was going, and a wood, cloth and fiberglass cowling behind it housed an alternator, regulator, and two storage batteries on the top of the tower. No vane was needed to guide this generator, and it ran quietly while the wind was steady, but squealed like an auto tire on a curve when the wind suddenly veered. In a 10 m.p.h. wind it produced ¼ horsepower, and at higher speeds it produced up to 2 horsepower. Meyer estimated then that it could be built for about $200 and offered to sell blocks of paper honeycomb to everyone who wanted to try. Hundreds of persons did, but no one knows how many succeeded.

The Windworkers went on to build 12-foot and 16-foot diameter solid-foil windmill rotors. R. Buckminster Fuller, who was impressed by the wind's power when he built his first Dymaxion house in 1927, has been their principal sponsor. "Wind power," he wrote, "permits humanity to participate in cosmic economics and evolutionary accommodation without in any way depleting or offending the great ecological regeneration of life on earth."

In the course of many experiments, Fuller and his associates found that the self-furling sails on the very old windmills still used on Greek islands were very efficient. The Wisconsinites consequently built 15-foot and 25-foot diameter sail windmills as well as machines with paper honeycomb airfoils. The publicity that the latter received, and Fuller's interest, attracted the attention of other wind researchers throughout the country, and the Wisconsin team has since worked closely with many others.

By 1975 Windworkers were selling plans and helping builders obtain materials and components for two types of windmills not seen in the United States before: One with a 12-foot honeycomb paper rotor, and a sail windmill with a rotor 25 feet in diameter.

The small wheel could be mounted on a 30-foot octahedron module tower, and had a 12-volt, 85-ampere alternator in its cowling. It was designed for 10.2-m.p.h. wind speed, and could draw up to 3,734 watts of energy from a 25-m.p.h. wind. The sail windmill had six triangular sails on a 42-foot tower, and transmitted power to a vertical drive shaft for whatever use one wanted to make of mechanical power on the ground. About half of that power could be put to use after the losses in the drive-train and alternator. But the cost of building almost any kind of wind-driven engine in one's own backyard was higher in 1976 than it had been earlier.

### ROTORS WITH RIMS

At the National Western Stock Show in Denver, a company organized in 1975 exhibited a new kind of American windmill that interested many well-informed people who were eager to use more of the wind's energy. American Wind Turbine, Inc., began manufacturing this machine in 1976 at a factory in Stillwater, Oklahoma. The smallest model it advertised had a turbine with a wind wheel 8 feet in diameter, and that turbine cost only $250. The company was also

*Builders foresee many uses for wind wheels like these in San Diego.*

prepared to produce much larger turbines and towers, and to sell kits of hardware to customers who wanted to assemble their own windmills.

The wind wheels resembled bicycle wheels. Thomas O. Chalk, an airplane mechanic at Ocala, Florida, had built a wheel 15 feet wide and put a rim around it. He had strung aluminum airfoils on the "spokes" between the hub and the rim of that wheel. Chalk's rotor weighed less than other many-bladed wind wheels, and caught more of the wind's energy. The surface speed of the rim was greater than that of the hub, and power could be drawn either from the rim or the hub of Chalk's wheel.

By using the rim of a large wheel to drive a smaller one, several kilowatts of electricity could be generated from the wind's energy. Chalk's invention was tested at Oklahoma State University, and Professor William Hughes became an officer of the American Wind Turbine Company when it was incorporated. The company suggested that its machines could be used to aerate ponds at fish hatcheries, light billboards along highways at night, and run air compressors at service stations for automobiles, as well as to pump water for cattle or to irrigate land.

Professor Hughes' confidence in this kind of wind-catcher's potentialities resulted in part from the progress made at his school and elsewhere in developing constant-frequency, variable speed (CFVS) electrical generators. "Some day," he told a subcommittee of the House of Representatives in Washington, "instead of making electricity from fuel, we will be making fuel in part from electricity."

Edmund L. Salter, a mechanical engineer in San Diego, built and began tests in 1975 of another combination of ideas. He placed three large rimmed rotors to intercept the wind around a small wheel constructed like a bicycle's wheel. All four wheels were in the same vertical plane, and in each one of three big ones there were three lightweight airfoils between the hub and the rim. From their tips the rim of the small central wheel was driven. Its hub was thus given sufficient speed for a shaft from it to drive a generator effectively.

Salter's prototype was designed to draw 7 kilowatts from a 28 m.p.h. wind. It was light enough for six big rotors to be supported by a single tower, and generate 42 kilowatts. The whole machine was designed for mass production, called the RD-7000, and a company was formed to manufacture and market it.

A newsletter edited by an engineer in San Diego for the Solar Energy Society listed the main features of Salter's new type of windmill as follows:

"Rim drive—provides a 5:1 speed ratio to drive the alternator without expensive, noisy, and unreliable gearboxes. Also, the rims will increase the efficiency of the rotors by as much as 20 per cent by preventing spillage of air at the blade tips. . . . Automatic high wind protection—allows the entire turbine to gradually turn away from high winds to prevent damage to rotors and structural members. This eliminates the need for very costly blade feathering mechanisms. . . . Polyurethane foam rotor blades with alloy steel spar tubes—combines an ultra-light airfoil with very strong, fatigue-resistant spars; each rotor blade assembly will weigh less than two pounds. A coating of tough, linear polyurethane will prevent damage from moisture, sunlight, temperature, and abrasive impact for the life of the unit. (No estimate given. Ed.) . . . Brushless alternator design with solid-state regulator—assures trouble-free operation. The only maintenance required will be bearing lubrication at five-year intervals. . . . Rotor and turntable bearings—automotive type tapered roller bearings for long life, low friction, and positive location. Again, a lubrication interval of five years. . . . System flexibility—capability for mounting multiple RD-7000 wind turbines on one specially designed tower for large installations."

Salter's company hoped soon to make an RD-7000 unit that could sell for about $2,700. A tower, batteries, and other equipment would bring the cost of a machine to generate 630 kilowatt hours a month up to about $5,000. The Swiss Elektro was the only machine then available in this power range, and it was priced at $6,500 for the turbine unit alone.

The RD-7000 could be a welcome development for anyone determined to generate his own electrical current for his home from the wind. But the initial cost would still be high until electrical energy could be stored more economically.

## ALL THINGS CONSIDERED

Modern management specialists favor a "systems approach" to all sorts of problems. If you want to build a better windmill that way you must study each part, the interfaces between it and other parts, and what can be done about each one. Social problems can be tackled the same way, and some people have begun doing it. They consider the environment, the sources of energy used and available, what people know and are accustomed to, how they are taught and can be influenced, and other factors. Whether your purpose is to improve the quality of life in some particular locality or to produce more food for mankind, this approach requires you to consider the environment. And when you do this, you discover the wind is a big factor in it.

The systems approach has led several groups of people eager to serve mankind to think about windmills. Community Technology, Inc., in Washington, D.C. and the New Alchemy Institute in Woods Hole, Massachusetts, are interesting examples of these modern do-gooders.

The community technologists would like to improve our cities. Carl Hess, who heads their non-profit organization, is a welder and former speech writer for Senator Barry Goldwater. Hess began by converting an abandoned gymnasium into a makeshift laboratory for volunteer craftsmen, engineers, teachers, and unemployed youngsters. To interest people with special skills and knowledge in this effort, Hess wrote an article for *The Washington Post* describing a utopian capital. There would be no dirty smoke from big utility plants in it because electricity would be produced from the wind. This might be done, he hoped, by little aerogenerators on roofs and fire escapes, and Savonius rotors on the side of the Washington Monument.

These city planners became so involved in other projects, however, that they postponed doing anything about the wind until they might have more time and money.

The New Alchemists in New England were even more ambitious. Their goal was to "create a greener, kindlier world," by producing food for its billions of people in new ways. John Todd, a restless young marine biologist, organized this institute with the help of friends in the big Woods Hole oceanographic and biological laboratories. The purity of the earth's land, water and air obviously would have to be restored to achieve their goal, and they gave windmills higher priority than seemed necessary to the community technologists.

The New Alchemy Institute obtained the use of a 10-acre farm on a Cape Cod hilltop as its laboratory for testing a variety of interesting ideas. There they set up a series of new kinds of wind-driven engines. Finding three of these machines in a clearing in the woods in the 1970s was almost like encountering Little Red Riding Hood's three bears, because one was very big, the next one was not so big, and the third one was a mere cub. All three of these contrivances were homemade and inexpensive.

The tower for the biggest one was a tall telephone pole supported by guy wires. On a windy summer day red cloth sails flapped around the tip of that pole as if they were birds tied to it by strings. The wind both shaped and wiggled in those traps. On the same tower the alchemists had previously tested solid airfoils made in a way that other research teams had suggested. The permanent components of this machine included the differential and other parts from an old car, and the whole shebang had cost its builders less than $450.

But they found it could be used either to pump water for organic gardening experiments or to charge storage batteries similar to those in golf carts. It was being used in 1974 to pump water into a solar heating system from which the water flowed into a tank for tropical fish under a plastic dome. When cooled the same water flowed downhill to a vegetable garden.

The medium-sized machine a few yards away was a pillar of Savonius rotors on a single vertical axle. Earle Barnhart, a bearded, suntanned young man in shorts, had helped erect it, and was still tinkering with it. This was clearly his favorite windmill. He pointed out that a man with a few tools could copy it and put one, two, three or more rotors on his pole to increase or decrease the amount of power he got.

But the smallest of these experimenters' three machines was the most fun to watch. It had a vertical rotor made out of a bicycle wheel, and its tower was just a fence post. To catch the wind with a bicycle wheel the alchemists had put thin strips of metal curved into airfoils between its wire spokes. This wheel was mounted on the far end of a horizontal bar, pivoted on the fencepost, and kept in the wind's path by a square rudder on the other end of the bar.

Someone had painted a bright smiling symbol of Old Sol on that rudder, and the blades in the bicycle wheel sparkled in the bright sunlight. But this contrivance was more than a pretty ornament. The wheel's original hub had been replaced by an Archer Dyna-Hub made in England. That hub contained an expensive but tiny generator, designed to produce enough current for a lamp on a bicycle. It was being used instead to keep a 6-volt battery charged on the fencepost.

Current from this miniature aerogenerator kept a small radio blaring out recorded music for the New Alchemists tending the fish in a pond under a hot plastic dome, weeding their organic garden, and selling nicely-printed accounts of their work to incredulous visitors. Was this a preview of agriculture in the future? Or was it a fantasy that the world would soon forget?

XV

# Academic Advocacy

My ventures are not in one bottom trusted.

MERCHANT OF VENICE, ACT ONE, SCENE 2

THE BOLDEST by far of this generation's advocates
of more use of the wind has been a retired U.S. Navy captain teaching
at the University of Massachusetts. William E. Heronemus was a
Wisconsin lad sent to Annapolis by Senator Robert M. LaFollette.
After serving on a destroyer in World War II, he went to the Massachusetts Institute of Technology, with no intention then of succeed-
ing Percy Thomas, M.I.T. '93, and Palmer Putnam, '23, as one of his
country's most effective pleaders for big wind-driven turbines to generate electricity.

At the Portsmouth, New Hampshire, Navy Yard Heronemus
ordered steel for the world's first nuclear submarine, the USS *Nauti-
lus*. More assignments to nuclear energy work under Admiral Hyman
G. Rickover followed, until in time the captain tired of the good old
navy way of doing things. He then went to work for United Aircraft
in Farmington, Connecticut, where engineers were studying the use
of hydrogen and fuel cells to produce and store electricity. An oppor-
tunity to teach lured him from there to a nondescript college hall in
Amherst, Massachusetts.

Both the pollution of the Connecticut River and plans to line it with nuclear power plants appalled Heronemus. So did the rising cost of electricity. He knew that windmills had "zero environmental impact," that safe nuclear reactors were hard to build, and that hydrogen research might soon solve the energy storage problem. He remembered seeing a boat sail directly into the wind at Annapolis because it had an underwater screw driven by a wind wheel whirling on its mast.

"I decided," he later told a newspaperman, "that there is indeed a conspiracy. The utilities and energy companies are going to get us hooked on nuclear power, then they're going to drive the price up wherever it will go. And who are they? They are the same friendly souls who are selling us coal and petroleum products."

In 1972 Heronemus prepared a paper for a joint meeting of mechanical and electrical engineers at Springfield, Massachusetts. In it he called three beliefs then prevalent in those professions mere myths. They were, first, that total windpower within the United States is insignificant; second, that average annual wind speeds of 30 m.p.h. are required for economical electrical power generation; and third, that "wind-generating power must be backed up one-to-one by other generating capacity."

Heronemus contended that wind-driven turbines might generate electrical energy economically on our country's coasts, Great Plains, and mountains, including those in Alaska. And he spoke of spending "megabucks" as casually as the nuclear physicists did. "Need we bother very much at all with fission and fusion," he asked, "if there are alternatives? Would there be some unacceptable stigma to our society were we to opt for an energy system whose science and technology would be very unsophisticated?"

This was still an heretical question in 1972. Oil companies were reassuring environmentalists by boasting of their success in preserving wildlife. Coal companies insisted that forests would soon rise on land that they had stripped. Heronemus was an obscure professor of engineering whose voice in the technical wilderness might never have been heard very far. But Mike Gravel, a United States Senator from Alaska, heard him and put his technical paper into *The Congressional Record*. That brought it to more public officials' and newspapermen's attention, and their interpretations of it intrigued state legislators and governors. Almost overnight, Professor Heronemus became a national celebrity acclaimed by people fearful of nuclear power plants.

Home handymen knew that the size of the current of air a rotor intercepted limited the energy you could get from the wind. Even with ideal airfoils you would need a tremendous rotor to generate enough kilowatts for an all-electric home. But Heronemus proposed to solve that problem by putting three or more rotors on a single tower. This might be similar to adding cylinders to an auto engine. But a huge number of wind-driven generators would be needed to equal the output of a single small nuclear plant. That might be done, he suggested, by building long chains of towers. The wind is always blowing somewhere, and if the chains of machines were long enough some units would always be producing power from it.

The towers could resemble those built for high-voltage transmission lines. They could be erected over open fields on the Great Plains. In densely populated areas a chain of towers might be put up over broad highways like New Jersey's Garden State Parkway. Even more kilowatts might be drawn from powerful ocean winds if towers were placed out on the continental shelf; platforms like those built for offshore oil-drilling rigs might support them there.

At first most nuclear engineers were appalled by the Massachusetts professor's thoughts. So were utility company executives. His figures and their own estimates showed that thousands of wind-driven generators would be needed to put a significant dent in the energy shortage. Developing, testing, and constructing the power plants that he envisioned would require huge investments in risky projects. Budget trimmers in timid government agencies wished that the former navy captain were writing science fiction rather than exciting laymen about such technological possibilities.

But Heronemus argued that, if big wind-driven turbines were mass produced in modern plants, the way many other things are today, they could be "a genuine bargain" by the 1980s.

## A HYDROGEN ECONOMY?

The energy crisis was really a greater economic than technical challenge. Scientists knew of many neglected sources of energy as well as numerous ways of conserving and storing energy. But like a child in a cafeteria, we have overloaded our trays with delicious deserts instead of making a wise selection of nourishing food. The staggering cost of reliance on fossil fuels may soon force us to turn to alternatives, and the wind's power is one of the most familiar of them.

Long before consumers realized the necessity for caution in using energy, noted scientists had predicted that mankind's principal fuel by the twenty-first century would be hydrogen rather than coal, oil, or natural gas. Aerogenerators, as well as other alternative energy generators that Heronemus has talked about, could be used to produce hydrogen by separating it from the oxygen in water.

Hydrogen is the simplest, lightest, and most abundant of all the chemical elements. The *Hindenburg* zeppelin fire in 1937 frightened nearly everyone who saw the pictures of those flames from hydrogen, but much more had been learned about handling it safely since then. Refineries are already producing significant quantities of it, and hydrogen was being transported safely both in tanks and pipelines. It was used to launch the Saturn rockets that carried the lunar explorers into space, and in the fuel cells that provided electrical energy in their spacecraft.

By the 1970s the Pratt & Whitney division of the United Aircraft Corp., had begun a $42 million program to develop megawatt fuel cells to help meet the peak loads on big thermal and nuclear power plants. "A fuel-cell plant to serve a community of 20,000 people," *Popular Science Monthly* reported, "could be assembled on a half-acre plot in the middle of town and no one would know it was there." That could make Heronemus' ideas much more enticing, both to persons concerned about the energy shortage and those who emphasized the preservation of a healthy environment.

Students at many universities began to take more interest in aeology than previous generations had, and were encouraged by faculty advisors to experiment with wind-driven devices. Both in states like Oklahoma, where the wind still came rustling across the plains, and in rugged coastal states like Oregon, research began that heartened other advocates of the use of windpower.

In Oklahoma the National Severe Storms Laboratory was using a 1,602-foot television tower to record data about the wind. Professor Karl H. Bergey at the state university in Norman used the new data thus obtained to solve engineering textbook equations. The answers he got suggested that it might be as feasible to use Oklahoma's winds to generate electrical power as to drain its oil fields.

Professor Bergey's students had built a hybrid electric car before gasoline prices rose appreciably. A wind-powered generator charged its batteries, but to extend the car's range more than 40 miles—and air-condition or heat the car—the students had to put a small internal combustion engine in it. Programs dependent on students tend to go by fits and starts, and Bergey wished he had more time to devote to aerogenerators. Within a few years, he predicted, "the virtues of windpower will be recognized and commercial production will follow rapidly."

In Oregon the Public Utility Directors' Association put up the money for an appraisal of the wind's potentialities by the state university in Corvallis. Professor E. Wendell Hewson, who taught atmospheric science, headed the project. His team included three mechanical engineers and a physicist. They found that the available meteorological data were inadequate, so they established more stations. With the additional data from them, the researchers soon identified several promising sites for wind-driven turbines near Portland and up the Columbia River valley.

When interviewed about this work, Professor Hewson likened windmill building today to the condition of the aviation industry in 1920. He recalled that if you asked then whether people could be flown across the continent more cheaply than trains could carry them, most engineers would have said no—but that's the way we go today.

Library browsers at most engineering schools discovered that more had been done recently about windpower in other countries than in the United States. The federal government and private industry here were betting almost wholly on nuclear power. Local legislators in some states recognized the demand for some other new source of energy sooner than national planners.

In Hawaii trade winds blow nine-tenths of the time in winter and half the time in the summer at average annual velocities from 10 to 15 m.p.h. at many sites. Imported fuel is expensive, and nuclear plants might drive tourists away. So the City and County of Honolulu sponsored a wind energy program at the University of Hawaii's Center for Engineering Research.

Its staff selected sites at high elevations on Oahu to study first. Meteorologists from the University of California's Lawrence Livermore Laboratory joined them, and Edward Teller, the physicist widely known as the father of the hydrogen bomb, became interested. "If wind velocity is to be useful anywhere, and it will be," he then declared, "it should be started in Hawaii where the chance of success is best."

## THE NATIONAL WORKSHOPS

As the cost of fuel continued to rise, people everywhere began to wonder why it should take so long to develop and evaluate windmills to generate electricity. In England a House of Commons Select Committee on Science and Technology asked Dr. Walter Marshall, the chief scientist of the country's Department of Energy. He replied at a public hearing attended by a record crowd, and *The New Scientist* summarized his explanation as follows:

"To begin with you have to analyse the data to see how much energy is available. Then you have to count the cost of the technology. Finally you have to do an energy accounting exercise. You cannot build windmills without using materials, labour, and energy. You have to do a careful study to see if you will gain energy or not."

That takes time, and prices change in every country.

Our National Science Foundation regarded the wind as a form of solar energy, and began by asking a panel of experts to examine solar energy as "a national resource" in 1972. They found that "the power potential in the winds over the continental U.S., Aleutian arc, and the eastern seaboard is about 10 billion kilowatts" and that the winds in many places are "remarkably repeatable and predictable."

World conferences on meteorology and energy, had not jolted our government into doing as much about using more of the wind as that report did. It prompted the science foundation and NASA to sponsor a national wind-utilization workshop in Washington in 1973. Windmills were no longer truly "mills," and purists defined the subject for discussion as "wind energy conversion systems." This was a fortunate phrase, because a fine acronym could be made out of it—"WECSes" rhymed with "nexus," and dignified the study of a source of power that most businessmen still tended to belittle. (It was a better term, at least, than "APE," the acronym for "assistant to the president of a company on energy problems.")

Many of the men mentioned in this book attended the WECS workshop. Beauchamp E. Smith showed a movie of the construction of the big power plant on Grandpa's Knob, and others reported on less costly experiments both here and abroad. Questions from the floor were friendly but brotherly, sharp and occasionally impertinent. On the third day the workshop participants agreed on recommendations which included a significant assertion:

"Compared to other sources such as nuclear, fusion, and geothermal, wind energy conversion systems are inherently simpler and could probably be made to be as cost effective and reliable as existing fossil and nuclear plants. The workshop participants believed that the cost of developing wind-driven power systems should be quite low compared to the costs of developing nuclear and other advanced systems, and that what was needed was a concerted and systematic attack on the economic and technical capabilities of wind energy systems."

This was a report that administrative agencies' spokesmen could quote to legislators reluctant to increase appropriations for further research and development of windmills. A five-year program was soon launched at NASA's Lewis Research Center, to solve the old problem of getting kilowatts out of the wind cheaply.

By 1975, when a second national workshop on wind energy conversion systems was held in Washington, many more engineers attended than had participated in the first one. Some represented new firms with scarcely any more physical assets than a postoffice box and a letterhead. Others represented corporations as wealthy as the Dutch East Indies Company gentlemen had been four centuries ago. The possibility of getting government grants or contracts for research also brought academic leaders from schools that needed help for both students and teachers.

Doubting Thomases still raised many questions about huge windmills: Would it be safe to drive on roads under whirling metal blades? Would birds and other wildlife be endangered? Wouldn't roads like the Skyline Drive in the beautiful Blue Ridge Mountains cease to be scenic if decked with windmills? Would windmills along sea shores be any more welcome than oil-drilling platforms? Wouldn't they interfere with yachting?

Engineers replied that aerogenerators over highways need be no more dangerous than high-voltage transmission lines, to people or other creatures. Secondary roads could still be charming. And offshore windmills might serve as aids to navigation. But common sense was not enough assurance for everyone, and eager researchers proposed to survey public opinion, collect gobs of data, and analyze it with computers.

Our federal government spent only about $200,000 on wind energy research in 1972. Three years later the newly formed Energy Research and Development Administration (ERDA) provided $7 million for about forty projects, and many Congressmen were willing to vote even larger appropriations for studies of new sources of energy.

Captain Heronemus, the Patrick Henry of the campaign gaining momentum, addressed both the first and second wind energy workshops. Men who had considered his ideas too grandiose in 1972 no longer did in 1975. Many had fallen into step with him, but he was still ahead of the majority in his thinking.

His team of researchers at the University of Massachusetts was analyzing models of systems in which both solar heat and the wind's energy might be used to conserve fuel. They expected soon to have a building especially devoted to such research. They hoped to develop a home heating system that would cost only a few thousand dollars to install and maintain for years. Millions of home builders, Heronemus predicted, would prefer such a system to big fuel bills.

Solar heating already was being tried in many states. Manufacturers were pushing new insulating materials and new devices to collect and horde heat, but buyers felt uncertain about which things and ways were best. Engineers at the Langley Research Center and elsewhere were testing some, and many more tests were being scheduled.

The problems of the solar heating enthusiasts and windpower advocates were similar: The initial cost of the equipment was high, and neither source of energy was as constant and convenient as the current from big central plants exhausting the stockpile of chemicals that nature had stored in the earth. But the demand for energy had goaded the government into doing more about it. In the future, ERDA announced soon after the 1975 workshop, solar electric systems would be "accorded high priority for long-term development along with fusion and the breeder reactor."

# Old and New Ideas

Sometimes the wind can come
And strike a house like a drum
Making the whole house shake
To bring the sleeper awake.

CHARLES NORMAN*

T O  R E T A I N a high standard of living, the United States needed a new source of energy by the 1970s that would be vast, harmless to the environment, and competitive with fossil fuels. Hundreds of engineers began to experiment with wind-driven devices, both in colossal laboratories and backyard shops. Some had the finest computers, scales, and other tools that money could buy; some made do, like the physicists a couple of centuries ago, with little more than love and string and sealing wax.

Most experimenters favored airfoils mounted to revolve vertically to the earth's surface. Others mounted the airfoils to whirl like merry-go-rounds, but no longer called them "horizontal rotors." Instead, they described them as "vertical-axis rotors" because the axis of rotation is perpendicular to both the earth's surface and the stream of wind intercepted.

Many beginners found a helpful source of information in Canada, where James Brace had endowed a small institute at McGill University to seek ways to help the poor people in arid lands produce

---

* From *Portents of the Air*, copyright © 1973 by Charles Norman. Reprinted by permission of the Bobbs-Merrill Co.

more food. Its staff had begun trying to do this in 1961 on a budget of less than $100,000 a year. In Barbados, one of the Windward Isles in the Caribbean, the Brace Institute's researchers had found an old stone tower that was still sturdy. On it they had mounted new metal wind wheels and aerogenerators to test them, and they had developed a prototype of a 10-horsepower turbine that was being used for irrigation.

Some of the most arid and least developed parts of the world are strewn now with rusting oil drums. Sophisticated technology has done too little to improve the plight of millions of people in these lands. Many have neither the skills nor materials required to build efficient machines, and are accustomed to using simpler aids to survival. We have four billion neighbors on our planet today, including hungry hordes who have made no use of the wind's energy. For them the Brace Research Institute has published complete directions for cutting oil drums in half to make Savonius rotors for windmills, and for others it has published more technical reports on various projects.

Until air-pollution alerts, electrical dim-outs, dry fuel tanks, and skyrocketing prices scared more engineers into action, few efforts had been made in the United States to generate mechanical power with the rotors that Savonius introduced in Finland in the 1920s. One of the first ideas that occurred to American experimenters was to stack two or three pairs of Savonius half-cylinders around a single vertical axle. More power could be gotten out of these contrivances, too, by making them out of lighter stuff than old oil drums.

At Woods Hole, Massachusetts, Earle Barnhart tried putting a small hump in one side of each of his half-cylinders to see if he could increase his machine's efficiency. At Crown Point, Indiana, John M. Thalmann hung very light, one-way flaps suspended from offset booms and mounted in pairs on a vertical shaft for the wind to spin. Others in other parts of the country tried other ideas. Michael Hackleman, who worked with a group called Earthmind in California, became so interested that he wrote a book about how to build a horizontal windmill. It describes a do-it-yourselfer's problems with aerogenerators realistically, and few of those problems can be solved as easily as many builders have hoped.

Engineers working for Canada's National Resource Council shared the interest of the Brace Institute's staff in helping people help themselves to the wind's energy, and made one of the most noteworthy contributions to that ancient art. Raj Rangi and Peter South unveiled a surprisingly new kind of rotor in the early 1970s that the wind whirled horizontally the way it had the Arabs' crude machines a thousand years ago. This one, however, more nearly resembled a machine that G. J. M. Darrieus built in France in the 1920s. It had not been further developed, and patents on it had expired. Newspapers everywhere published pictures of the new Darrieus type of windmill in Canada because it looked more like a big advertisement for a kitchen mixer than a working windmill. It ran so fast and easily that reporters dubbed it an eggbeater.

Vertical propeller-type rotors were clearly more efficient and better understood than vertical-axis machines. But the Canadians got 900 watts from a generator driven by a rotor 15 feet wide. The sails of their pilot model were thin metal airfoils, bent into big bows and attached at both ends to a vertical axle that was held upright by guy wires. The generator was on the ground directly underneath those novel sails.

This eggbeater started reluctantly, but when an engineer gave it a shove by hand it soon spun extremely fast. In a 15-m.p.h. wind, the blades revolved 170 times a minute and the peripheral speed of the outermost part of the bows was six times the wind's velocity. That impressed other designers of aerogenerators.

The nine-tenths of a kilowatt produced by the eggbeater was not much energy by today's standards, but it was enough to be mighty useful in the dark Indian villages and at military bases in the Far North. Transmission lines had not been extended to all parts of Canada, or to many other inhabited parts of the world, and the Canadian government encouraged further development of the power plant Ranji and South had built. Some engineers, however, thought it might be difficult to manufacture bow-shaped airfoils economically, and no one was prepared to estimate quickly the cost of mass producing and distributing so unusual a windmill.

In the Philippines the International Rice Research Institute mounted a two-bladed rotor similar to the Canadian machine on a pick-up truck, and found that it would not start itself even in wind speeds of 57 m.p.h. But in the United States experimenters soon found ways to put "self-starters" on bow-shaped Darrieus rotors.

### NASA'S FIRST WINDMILL

Dr. Emil J. Steinhardt and his colleagues at West Virginia University raised questions that led to the first significant study in the United States for many years of horizontal windmills. The professors thought it was shameful that poor people in the Appalachian mountains had no electric lights in their homes, and went to the Langley Research Center to consult its staff about what could be done.

There they met Dr. John D. Buckley, a forthright muscular engineer working on plans for a spacecraft to land on Mars. He had paid no attention to windmills since becoming absorbed in questions of flight and interplanetary exploration. But he and several of his friends at the laboratory were instantly sympathetic. They had studied engineering at Clemson University in South Carolina at a time when students had to construct whatever apparatus they needed for their experiments with whatever materials and tools they could mooch or borrow. They knew, too, what life was like for many families in Appalachia.

Dr. Buckley recalled having read a magazine article about the egg-beater Ranji and South were testing in Canada. When he described it to his friends, they thought it might be fun to build something like it. In spare time they searched scrap piles and junkyards, and soon had everything they needed.

Then, before diligent accountants and safety inspectors quite realized what was going on, a strange device began to generate electricity on the roof of a laboratory amidst the big wind tunnels, hydrogen tanks, and computers used for aeronautical and astronautical research. It worked like the one in Canada but differed in several respects.

Instead of metal, the two bow-shaped airfoils were made out of balsa wood covered with fiberglass. Each one was 6 inches wide, and less than an inch thick. The airfoils were sanded smooth and bent outward in the center to form a rotor about 14 feet wide. They were fastened to mounting plates about 13½ feet apart on an aluminum tube 4 inches in diameter. This mast rose from a 4-foot high pedestal, 4 feet wide at its square base.

In that open base there was a small alternator to generate elec-

*NASA's vertical-axis—"eggbeater" type—windmill at Langley Research Center.*

tricity and a disk brake and motor. An electric starter "cranked" the rotor up to speed on this machine. The builders did not expect to draw much current from the wind with so hastily built a machine, but their initial measurements of its output were as encouraging as those reported for the eggbeater in Canada.

"We're just in stage one," Dr. Buckley insisted. "We've got a lot to learn." He did not expect this machine to compete with big power plants. But he thought it might be manufactured cheaply, and one with a much larger rotor might generate a few kilowatts for enough hours at some sites in Appalachia to brighten life in humble homes. Few visitors to the NASA research center noticed it, and the management was tolerant of engineers' hobbies. If a CBS television crew had not shown up to film some of the other work under way one day, the builders of this windmill might have gone on quietly testing and tinkering with their plaything much longer in spare moments.

The wind-driven bows caught a cameraman's eye, and when a picture of the Langley windmill appeared on TV screens across the nation, the front office received hundreds of inquiries about it. Some came from home handymen and some from influential Congressmen and officials at NASA's headquarters in Washington. So Langley's public affairs officers had photos taken and wrote a news release about the new machine.

Safety officers became worried lest someone stick his head between the bowed-out blades and be knocked out. Auditors wanted records of costs. Aeronautical engineers wanted the machine moved into a wind tunnel to analyze its performance more thoroughly. More help was needed to handle the paper work. Acquiring enough data to satisfy skeptical experts and developing the machine further ceased to be a mere pastime. Vertical rotors may always be the most efficient way to extract energy from wind, but a vertical-axis rotor that needs no guidance may be the most practical solution to the energy problem in some situations.

### THE HYBRID AT SANDIA

The next eggbeater to attract national attention was in the Atomic Energy Commission's Sandia Laboratories at Albuquerque, New Mexico. The builders dubbed it a VAWT, an acronym for "vertical axis wind turbine." It was about the same size as the Canadian and Langley machines, but was constructed differently and had three blades. It started itself without being pushed manually or electrically, and could draw an impressive number of kilowatts from an 8-to-10 m.p.h. wind.

The VAWT started easily because it had two "buckets" on the same vertical axle with the eggbeater blades. They contained curved fins that functioned like Savonius half cylinders. These fins were oriented in different directions and caught enough wind to give the impetus to the big blades that they required. The three eggbeater blades were attached to these buckets, and each one had three parts: both ends were straight, but the middle section was a "troposkien." That is the technical term for the shape a child's jumping rope assumes when being used, and it was chosen to minimize the bending stresses.

Three men in Sandia's Aerodynamics Projects Department, Bennie Blackwell, Louis Feltz, and Randall Maydew, built this machine. They pointed out that it would be simpler and easier to make and ship big windmill blades made the way they had devised. This could reduce the cost of installing small power plants. And when the Energy Research and Development Administration (ERDA) took over the Atomic Energy Commission's laboratories it applied for a patent on the improvements made at Sandia in the eggbeater.

"Our initial effort was to provide electricity for remote locations," Blackwell told *Popular Science Monthly*. "But now we're hoping to present a technological base to the ERDA so that the Darrieus turbine can be considered as an alternate concept for producing large quantities of electricity to feed into the existing power grid. We think this turbine has the potential to do it more economically than the propeller system."

The first eggbeaters were not as pretty as some of the older types of windmills, but architects in Denmark have shown how they might be equally attractive when designed to be mounted on the roof of a modern building, or on a slim tower in an open field. The Danes have even considered a plan to stack possibly six of these rotors in a very tall, open tower which could be as beautiful and animated as any windmill ever built.

Americans have also recalled other often forgotten ideas (see Chapter III) since taking a new "hard look" at horizontal rotors. The staff of

*Sandia's VAWT interested ERDA in vertical-axis wind turbine research.*

181 ैॐ

*Two proposed Danish designs for vertical-axis wind turbine towers.*

the Advanced Concepts Division at Science Applications, Inc., in La Jolla, California, has thought of using sunshine to help whirl Savonius rotors. Suppose that a circular wall were built around a hot part of a desert. The enclosed surface might be covered or treated in some new way to make the air even hotter there. That air would rise, and cooler air would rush in through slots in the walls. There could be Savonius rotors in those slots—and the air racing through them would make the blades revolve faster. Such a power plant would not surprise the ghost of an ancient Arab as much as a line of automobiles waiting for a drink of the fuel now pumped from beneath the hot sand that he had trod.

Time after time advanced thinkers have thought of using ducts, funnels, or chimneys to gain more power from turbines driven by the wind. Some have suggested that aerodynamicists strive to create miniature, controllable tornadoes to whirl rotors. Their ideas are reminiscent of the French plan to collect hot air on the Algerian desert and shoot it up an immense chimney to drive a turbine in the cool air on top of a mountain. That was never done, but ducts are an idea that may still be revived.

The National Science Foundation has supported a study of another plan that reminded old-timers of the New Jersey utility companies' idea of putting Magnus cylinders on railroad flat cars on a circular track. At the University of Montana professors have proposed to hang aerogenerators on a round or oval skyway several miles long. Computers constantly informed of the direction and velocity of the approaching wind would shift the generators around to keep each one where it would drain the most energy from the wind.

The wind wheels might resemble airplane propellers, and their combined output in Montana's wide open spaces would be greater and steadier than that of any single machine. Cattle might graze undisturbed under that power plant held high in the sky by cables between tall towers. Engineers might enjoy seeing such a plant run itself with minimal human supervision as much as people once enjoyed watching a miller at work on a Dutch tower mill's stage.

But it would certainly alter Montana's scenic beauty. "You are not going to get the ecology movement to hold still for a gang of windmills hanging from a balloon or a wire," an Oregon public utility executive warned the Montana professors. "Anything that sticks up and attracts attention is going to be attacked by the ecology people."

## THE STORAGE QUESTION

A greater obstacle than the ecologists to realizing engineers' dreams has been the cost of storing kilowatts. The wind will not produce them like a utility company the instant that you may want them. Storage batteries have been used with most small aerogenerators, and they cost more than barnyard water tanks. Utility companies would welcome a cheap way to stockpile electricity, because the demand for it fluctuates like the weather.

Archeologists suspect that a battery may be as old a device as a windmill. In the outskirts of Baghdad and other ancient cities, they have found bits of metal and vases that magicians may once have used to make batteries. If so, Alessandro Volta's research less than two centuries ago might have explained some of those magicians' incredible feats.

In the early 1900s nearly everyone supposed that batteries would soon be greatly improved. When steam-driven dynamos first generated direct current for some small towns, the steam engines had to be shut down to clean the boilers periodically. Big banks of glass storage batteries sometimes stood alongside the dynamos then to keep a few lights lit. Some engineers felt so certain that this way of storing electricity would soon become economical that they questioned the wisdom of building big central power plants and long transmission lines for alternating current.

Better batteries are available today for automobiles, flashlights, electric razors, hedge shears, and other things, but those suitable for windmills are still expensive. Lead-calcium cells are being used instead of lead-acid batteries with French Aerowatt aerogenerators. Some other new type of battery, possibly nickel-iron or nickel-zinc, may reduce the cost of converting mechanical energy into chemical energy to save it until you want to turn it into electrical energy. But every conversion involves some losses of energy and requires equipment that adds to the storage cost.

Fuel cells could be the answer. Fuel cells differ from conventional, self-contained batteries in having the chemicals that they use to generate electricity fed into them continuously from an outside fuel supply.

Some engineers, however, think the best way to store mechanical energy will always be by mechanical rather than chemical means.

Energy drawn from the wind, for instance, could be used to compress air for storage. We use compressed air effectively in every filling station now. Some day air may be compressed for storage in big caves or inflatable tanks under the sea. If this were done compressed air might be permitted to rush out clean and fresh in measured amounts to run turbines whenever the wind stopped blowing.

A flywheel is another old device that may be made more efficient. The flat disk that a potter molded his clay on in Omar Khayyám's day was a flywheel. When an engine delivers more energy than is needed right then—the way windmills often do—a flywheel can absorb some of that energy by running faster. Later the flywheel will continue to revolve for a time from its own momentum.

Flywheels are used to make both large and small engines run more smoothly. Those now in widespread use will fly apart when greatly overloaded, but at Johns Hopkins University and elsewhere researchers are working on "superflywheels." They would be both stronger and lighter than today's flywheels because they would be made in new ways out of new composite materials. Fine filaments of very strong stuff can be put in those materials, and possibly arranged like the hairs in a rotary brush, to retain more energy for longer periods than flywheels do today. But if and when superflywheels are made they will probably be expensive at first.

The designers of energy conversion systems for spacecraft did not worry about possible costs of mass producing new devices. They were concerned only with getting unprecedented jobs done to meet deadlines. But they became cost conscious in the 1970s when asked to think about storage systems for use with windmills.

They began by considering batteries, compressed air, flywheels, and hydrogen, but some soon favored a relatively new idea: electrically rechargeable "redox" (oxidation reduction) flow cells. Energy would be stored chemically in those cells, but the materials in the electrodes and fluids might not be rare or costly, and the tanks might be huge and simply constructed. A NASA Technical Memorandum in 1974 reviewed the status of this technique then: "The concept of the electrically rechargeable redox cell has been described mainly from a theoretical standpoint. Preliminary results from laboratory experiments point up problem areas in this concept. Thus far the contentions of high overall electrical efficiency as well as deep depths of discharge have been supported by the data. Proper membrane materials have yet to be found and more work must be done to select the best redox couples. Cost studies can then be undertaken to establish estimated costs (fixed and operating) for systems of different sizes."

Development of redox cells could obviously take considerable time. Science reporting has long been a treacherous trade, vulnerable to wishful thinking, and the most productive research often is, too.

# Small Wind-Catchers

Had it not been for the abundance of fossil fuels—coal, oil, and natural gas—we might today have a Solar Energy Economy as effective and efficient as our "Fossil Fuel Economy."

NATIONAL PETROLEUM COUNCIL REPORT

MANY AMERICAN FARMERS built their own windmills when the twentieth century began. A Texan, Professor Webb, conceded that Nebraska had "the greatest aggregation of home-made windmills known to history." Professor Erwin Hinckley Barbour (whose interest in "merry-go-rounds" and rhinoceros bones was mentioned in Chapter III) described those windmills in a bulletin that he wrote for Nebraska's agricultural experiment station. He reported that the three principal types were known locally as Mock Turbines, Jumbos, and Battle-Axes.

The Mock Turbines were the simplest ones. Their wheels were rough-and-ready imitations of the fans on the factory-made windmills, and the towers often were poles cut from trees. They were most popular near Grand Island, a growing trading center in the shallow Platte valley. Friederich Ernstmeyer made one out of cottonwood poles, a plank that cost him only 12 cents, and parts he salvaged from a mowing machine and cultivator. Frederick Mathiesen invested nearly five dollars in parts for a fancier model. It had a six-bladed wooden fan supported over a water tank by four locust poles, and pumped water for fifty head of cattle.

An ingenious chap, it seemed, could make a windmill powerful enough to pump some water with whatever happened to be handy.

Many men used the hubs and axles of old wagon wheels. More often than not the wind in Nebraska came from the north or south, and a respectable bit of energy could be obtained from it without having to swing a rotor to the east or west. But some builders attached rudders to their rotors to hold them in the wind, and some made brakes by adding a lever that they could pull from the ground with a wire.

The Jumbo machines were the least vulnerable to the wind, and Barbour considered them the least efficient of the windmills that he studied. The rotor of a Jumbo resembled an overshot water wheel constructed by attaching paddles to a horizontal axle. The axle usually extended across the top of a plain wooden box, so that the wind would strike each paddle when it was above the box but not retard it when that paddle dipped into the box. One Nebraskan raked a heap of tomato cans into a straw fire to melt the solder, then flattened out several hundred pieces of tin from them to make paddles for his Jumbo.

You could hammer a Jumbo together without an expert's help, and there were as many sizes of them as of elephants. The professor found baby Jumbos, medium-sized, and giant Jumbos. The biggest ones were called "go-devils." One that Edward Murphy tested in Kansas (as noted in Chapter X) was as powerful as some of the most expensive factory-built windmills. Near Kearney, Nebraska, the midway point between Boston and San Francisco, a Jumbo that cost its builder only $1.50 irrigated his strawberry patch and some small fruit trees; and Jumbos were especially popular in eastern Nebraska.

In the western counties, farmers tended to prefer Battle-Axes. They were so called because the sails were mounted on the far ends of long arms and seemed to chop a strong wind in opposite directions like men swinging double-edged axes in a bloody battle. Those sails might be any of several shapes, and made out of wood or almost any other tough material.

Jacob Geiss had a typical Battle-Axe on his place near Grand Island. The wooden sails of its 12-foot rotor were 3 feet square and whirled a heavy wooden axle 8 feet long. That windmill cost Geiss less than $12 and pumped water from a shallow well for 125 head of cattle. Farther west Barbour found two well-built Battle-Axes on the 1,600-acre Matthew Wilson property, helping several shop-made windmills water several hundred head of cattle and a garden patch.

Above is a typical Battle-Axe windmill, used around 1898 to pump water for stock on a farm near Overton, Nebraska. Below, an unusual homemade windmill that was put up near Ashland, Nebraska, around the same time.

Some owners took great pride in their Battle-Axes. One builder near Cozad painted the tower white and the fan red, and kept it well oiled to pump water into a big tank for his home, barn, and lawn. Southwest of Gothenburg big Battle-Axes helped shop-made windmills irrigate a thriving 16-acre orchard and another one newly started. Even when built with new material, a homemade Battle-Axe seldom cost more than $40 or $50.

When *The Lincoln State Journal* offered a prize for the best homemade windmill in Nebraska it went to a novel two-bladed machine perched on the roof of a shed near Ashland. Elmer Jasperson, who built it, had spent exactly $11 for materials and used that windmill to run a two-hole corn sheller, a feed grinder, and a grindstone in his shop. Barbour described it as follows: "It is a sort of adjustable Battle-Axe, the fans turning upon the short arms to throw it in and out of gear. When thrown out of gear the two great semi-circular fans make a ten-foot circle with the edge to the wind; accordingly it remains stationary. However when the two fans are slightly oblique, then the force of the wind is felt, and the mill starts."

Power plants were then becoming more complex and difficult to duplicate outside a factory. A machine that worked well was usually copied by its builder's neighbors, but Jasperson's prize-winner was not. Men built windmills throughout the Great Plains a few score years ago mainly because it was easy to do and cost little. "Let them be built cheaply," Barbour advised his readers, "or not at all."

Those mills were not intended to last long. The Nebraska State Historical Society has preserved copies of Professor Barbour's accounts of them, but does not know of a single Mock Turbine, Jumbo, or Battle-Axe still upright.

### FRINGE BENEFITS

With a few tools almost anyone can still build a windmill to run a pump. Published plans, blueprints, and directions have made it easier than ever. Today, however, it may be cheaper to repair an old windmill than to start from scratch. New Mexico State University's faculty estimated recently that 70,000 old windmills in the United States could be made operable again, and offered a special course in how to do it.

Although professional mechanics insist that repairing windmills is a hard way to make a living, the fringe benefits make it a rewarding way to spend spare time. A windmill may give a person a sense of permanence and security that is hard to recapture otherwise. In Arizona, Ida Goff wished she had a windmill like one she remembered in Kansas. Her husband found an old fan wheel he could repair, and her son built a 30-foot tower for it out of scrap iron. "There aren't many things I want," Mrs. Goff said when they put it up, "but this is one of them. It'll make a great conversation piece."

In California, "Doc" Dachtler fixed up an old windmill to pump water for a cabin in the Sierra foothills. Neighbors asked him for help with other old machines, and rewarded him with fresh milk, cream, and home-grown vegetables. The Dachtlers were one of many young couples who preferred a rural region to a chaotic city in the 1960s.

Like Aesop's country mouse, they thought "Rather beans and bacon in peace than cakes and ale in fear." Urban delights have changed in this century, and many migrants to the backwoods missed electrical outlets more than plumbing. So some turned to the wind for energy. But building a windmill to produce electricity differs as much from making one to produce mechanical energy as building a motor boat does from making a raft—and often can best be done with the help of components from an old one.

In Colorado, Winnie Red Rocker especially missed stereo music. To produce current for it he made an aerogenerator by carving a propeller out of a piece of redwood to connect to some parts from an old car and an old Wincharger. That was "pretty easy," Winnie reported happily in a newsletter called *Alternative Sources of Energy*. He had not expected much current from a homemade machine, but others did.

So many people soon wanted to make aerogenerators that parts from decrepit factory-built machines became scarce and expensive in some areas. James Harrison of Danby, Vermont, got a nearby iron-works to make a four-bladed rotor for a tiny generator he built. Figuring out specifications, gear ratios, and other details he admitted, was "quite a lot of slide-rule work."

For some men, this work has been an additional source of satisfaction when they succeeded. Earle Rich, an electrical engineer, put a three-bladed wheel 20 feet in diameter on a 60-foot tower at Scarborough, Maine, for a generator capable of delivering 15 kilowatts from a 35-m.p.h. wind. Even though Rich used surplus parts, that machine cost him $2,200.

Newspapers published glowing accounts of many homemade machines, and some builders went into the windmill business. Jim Sencenbaugh, a young man in Palo Alto, California, was one of them. Lt. Robert Reines, who was born in Los Alamos where his father worked on the atomic bomb, got a $25,000 grant from the Air Force to experiment with solar energy for a home, and used three windmills to generate electricity for it. Successful windmill experimenters often were widely acclaimed, and those who failed were ignored like losers at Las Vegas.

"Be prepared," the Boy Scout motto, is a good one for ambitious home handymen. Many Americans had no idea how many kilowatts their families routinely use until utility bills began to go sky high. Although people notice the difference in the wattage of light bulbs, many have bought appliances without knowing how much current they would require. Even a modest home may now have a 12,200-watt kitchen range, a 4,474-watt water heater, a 630-watt vacuum cleaner, a 615-watt refrigerator-freezer, a 512-watt washing machine, and a 330-watt television set.

Those numbers must be multiplied by the number of hours that each appliance is used to determine how much current it will draw annually in kilowatt hours. Some appliances are run only seasonally; and some turn themselves on and off automatically. Even a 2-watt electric clock will take 17 kilowatt hours a year. Utility companies have distributed estimates of the annual consumption of many appliances, but those are national averages, and your family may be accustomed to consuming much more. An all-electric home may require more than 20,000 kilowatt hours a year, and today an aerogenerator to produce that much is a luxury few people can afford.

Even shopping for a ready-made aerogenerator was a challenge to an engineer in the 1970s. The rating a manufacturer gives a machine tells you only how many watts it can draw from a wind blowing neither too fast nor too slow. A 100-watt generator, for example, will only extract 100 watts of energy from wind within a certain range of velocities. Anyone who wants a wind generator should remember this, and should know approximately how many hours a year wind is likely to have a certain speed at the place where he expects to put his machine up.

### WINDMILLS FROM AUSTRALIA

Henry Clews was better qualified to build a wind generator than most men when he moved into an abandoned village in Maine once called Happytown. Clews was an aeronautical engineer who knew how to use tools and had taught science in a Portland high school. He found a suitable foundation for a home on the main street of the old town, and built a fine house for himself, his wife, and two children. But the Bangor hydroelectric company had no transmission line near it.

Mrs. Clews did not mind cooking on a wood-burning stove or doing without some other electrical appliances but she objected to pretending it was 1740 when it was not. The power company wanted so much for building a line and serving Clews' home that he decided to construct his own home-lighting plant. But after reviewing his textbooks, studying the wind and all the available plans he could find, he concluded that he would rather buy than build a wind generator.

Unfortunately no machine being manufactured in the United States would meet Clews' modest requirements. But he found that a 2,000-watt Dunlite "brushless windplant" being made in Australia might, and ordered one. When the machine arrived in Maine, it resembled a giant Erector Set. No one could have assembled it and erected the tower without help, but Clews' family and neighbors helped him get it up and running. The engine cost $1,475, the tower $315, storage batteries $475, and additional equipment and shipping costs ran the total bill up to $2,790. In the end, it would have been cheaper to have had the utility company build a transmission line.

Clews mounted the 40-foot galvanized steel tower on three concrete pylons sunk three feet in the ground. The rotor had three long rounded and tapered blades and drove a 115-volt generator. It charged a row of lead-acid batteries with a storage capacity of 130 ampere hours. Clews wired his home so that he could use either alternating or direct current, and had a 250-watt inverter to supply whichever kind he wanted for a particular purpose.

193 ॐ

The wind plant so delighted Clews that he went into the business of selling imported aerogenerators. Neil Welliver, an artist living in Maine, was one of his first customers. Since then Clews and Welliver have each bought a second foreign-made machine to have more electrical power in their homes.

"It was not really the money for us," Mrs. Welliver explained. "We're really doing this out of perversity."

Henry Clews also wrote and sold a booklet full of practical advice for anyone thinking of building or buying an aerogenerator. In it he emphasized the importance of measuring the wind's speed at the precise place where the rotor will intercept it and checking one's findings with the records available at the nearest Weather Bureau. Sailors often become quite good at guessing the wind's speed, but few landlubbers are. Clews suggested taking a reading every day at the same time, preferably in the afternoon, as long as possible.

"Of course," he wrote, "this may mean climbing a 50-foot tree every afternoon at 3 P.M., but the exercise will do you good!"

A conscientious Australian lady, Dorothy M. Gulson, was largely responsible for sales of her country's windmills in the United States. When she finished high school her father had told her she would be entering his business the next year and, as she says, "in 1913 a dutiful child did not argue with her father." Ms. Gulson became managing director of the Quirk's Victory Light Company in New South Wales, and Quirk's sold Dunlites overseas.

When I inquired about buying one, the company put some good questions to me before quoting a price. How big was my home, how was it lit, and had I already tried a direct current generator with batteries? What electrical appliances did I have and anticipate buying? How many watts would each require and how much would it be used? Would I need alternating current for some of them? How much of the time does the wind blow 20 m.p.h. where I live, and does it ever exceed 80 m.p.h.? How high was my house and the trees and other obstructions near it?

Those are good questions for everyone to ask who is thinking of buying or building a wind-driven power plant.

Paul Taylor, a dentist in Sonoma County, California, bought a

Dunlite and put ten 12-volt truck batteries, a voltage regulator, and a control panel in a shed beneath a 25-foot tower. The Taylors found the free current from the wind and this storage system adequate for lights and a few small appliances in their home. When the wind was really blowing outside, they could use its power to run the washing machine. But after a year's test of its usefulness to them, they kept their home tied to a utility company's lines and reported, "We're very careful with our electricity."

John G. Schauber, Sr., a Dunlite owner and dealer at New Castle, Delaware, thinks many more Americans would have Australian aerogenerators now if the dollar had not been devalued.

### WINDMILLS FROM EUROPE

Henry Clews chose a $4,000 Elektro G. m. b. H. wind generator manufactured in Switzerland for his second machine. It was being used in ski resorts in the Alps and was said to be the largest one anyone could buy at the time. At Port Richmond, California, a young architect and his colleagues building energy-saving homes, also chose the Swiss machine to serve them. Robert Dietrich told reporters, "We have so damned much wind out here it seemed a shame not to use it." That aerogenerator drew 6 kilowatts from the wind whenever it blew 20 m.p.h. It was mounted on a 35-foot tower within sight of San Francisco's skyscrapers, and each home would have standby power from the Pacific Gas and Electric Co.

In Germany a more modern pumping windmill was being manufactured than any American factory was producing. This Lubing machine had a tower made out of plastic tubing and a rotor with three aerodynamically profiled blades. Dr. Harry P. Bugben, a retired engineer associated with the Brace Research Institute, promoted sales of the Lubing machine in Canada.

In France a small Aerowatt machine is being built by a firm that gave up constructing huge generators after one lost a blade in 1960. The Aerowatt still being produced is a two-bladed machine small enough to be mounted on a rooftop or a large boat. One size generated 24 watts and a larger model produced more watts to charge storage batteries. The blades revolved in a 7-m.p.h. wind and reached

*NASA chose a tiny French Aerowatt to study at Plum Brook, Ohio.*

their top speed and electricity output at 15 m.p.h. The Aerowatt's blades could be coated with protective materials to protect them from ice and sandstorms. The machines have withstood 150-m.p.h. winds. This has made the Aerowatt an attractive machine for use at sea or in places far from utility lines. Several importers have sold Aerowatts in the United States.

The most powerful aerogenerator in use anywhere in the world in 1973 was a 70-kilowatt experimental machine on the West German island of Sylt, and it was also one of the most interesting ever built. A group of scientists in Geneva headed by Walter Schoenball had conceived it and they called it Noah, but whether their machine could be an ark to save mankind from an energy crisis was still speculative.

Schoenball was a lawyer who had become a population expert for the United Nations. His team included both French and German engineers, and they used their own money to construct the prototype

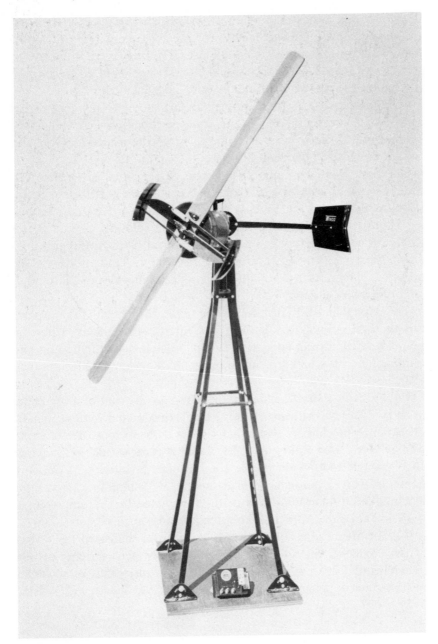

*Winchargers have proven small machines' usefulness.*

at Sylt. It had two 36-foot-wide rotors with five especially designed airfoil blades in each one. Those rotors revolved in opposite directions, and the difference in speed between them was sufficient to generate electricity without gears. Doing this was reported to simplify and reduce the cost of manufacturing an aerogenerator.

"Old-fashioned windmills trapped an average of 10 per cent of the wind and a maximum of 30 per cent," Schoenball told reporters. "Our windmill makes use of 70 per cent of the available wind and even as much as 80 per cent."

The first Noah cost about $190,000, but Schoenball believed it could be reproduced for less than $46,000 if made in sufficient quantities. Assuming his machine would last about twenty years, the cost per kilowatt hour might be only 2.7 cents a kilowatt at many sites regardless of what happened to the price of oil. The prototype would possibly produce enough current for five homes, and its developers prepared blueprints for 30-kilowatt and 230-kilowatt machines that might be built similarly.

The Mongolian Socialist Republic, much to Schoenball's surprise, was the first country to manifest any interest in his team's innovations. Americans, too, later tried to buy Noahs, but none was up and running here yet when our country began celebrating its bicentennial.

"The dream of a simple machine whirring on the roof of a log cabin in the wilderness, with people inside enjoying modern conveniences, turns out to be elusive," J. Baldwin wrote with obvious reluctance in The Whole Earth Epilog in 1974. Commercially available machines large enough to provide power for business as usual in a modern home are very expensive, he explained, and "rebuilding an antique machine often takes more money than predicted by the time you get it up on a tower and provide the batteries and controls."

But the dream may soon become less elusive, thanks to the revival of zeal for using the wind's energy and government support of new research and development programs. Windmillology may soon change faster than ever before.

# XVIII

# *The Windmill's Rejuvenation*

Madame, bear in mind
That princes govern all things—save the wind.
VICTOR HUGO, "THE INFANTA'S DREAM"

''AMERICA, although but a child of yesterday, has already given proof of its genius," Thomas Jefferson wrote in 1781.
Our country gave more proof in its second century, and can produce
even more by the tricentennial year. If we try, we can, for instance,
show people everywhere how to use the earth's many resources more
wisely.

American women rediscovered a "solar-powered drier" called a
clothesline when electric bills began to soar. Engineers simultaneously
renewed attempts to generate electricity from a "solar-powered air
stream"—the wind. Watching windmills consequently has become an
increasingly exciting sport. Artists, writers, environmentalists, and
ecologists will especially welcome a rejuvenation of the old dame
many of us still call a windmill, whatever engineers may call her.

Dutch and English post and smock mills have become historic treasures. The rotors on the new vertical-axis wind turbines revolve horizontally like those on windmills once called "merry-go-rounds." The
American windmills invented a century ago are still useful, but future
wind-catchers will look more like airplanes and spacecraft than like
the nineteenth century's windmills.

When 40 teams from 33 universities took part recently in a Student Competition for Relevant Engineering—called SCORE, of
course—at Albuquerque, many of them exhibited devices for captur-
ing solar heat and wind. Their elders were simultaneously testing new

rotors in several states; Rhode Island had a vertical-axis wind turbine on the roof of a laboratory near the statehouse in Providence. And computers in other shops were spewing out numerical forecasts of the performance of many more novel machines for the wind to drive.

Ferris wheels built 80 years ago were 300 feet in diameter. Bigger wheels could be built for windmills now, but the towers to support them would cost so much that it has not been done. Small rotors have been studied more intensely, and researchers have sought favorable sites for them. At Annapolis, Maryland, Peter R. Payne has suggested that the powerful wind currents at high altitudes be intercepted by an aircraft flown and tethered like a kite.

Windmill watchers may not see that done very soon, but those wise to the ways of the world will watch what happens on the Ohio shore of Lake Erie. The space engineers at the Lewis Research Center in Cleveland have put up a new wind turbine there, and will use it to test new concepts and components for wind energy conversion systems (WECSes) for the Energy Research and Development Administration (ERDA).

When astronauts explore the surface of Mars they may want energy from its wind, but a more immediate need accounted for NASA's interest in wind turbines. It resulted from a problem that arose on a speck of land called Culebra about eighteen miles from the mainland of Puerto Rico. That tiny island had so few natural resources that the U.S. Navy had used it for target practice. But people live there, and they wanted a power plant.

Meteorologists have suspected that one flap of a sea gull's wings at precisely the right instant and position in the sky may affect the weather in distant places. A question from the Governor of Puerto Rico resulted in a comparable sequence of current events. He asked the National Science Foundation what kind of power plant might best be built on Culebra. It bounced the question to NASA's Lewis Research Center because its staff was noted for its expertise about engines, materials, lubricants, and related subjects.

Joseph M. Savino, a philosophical engineer with a sense of humor, and other Lewis specialists found the challenge from Culebra intriguing. The wind was known to be both strong and reliable there, and Savino went to Culebra to tramp the little island's hills for a couple of days looking for sites for windmills. Lewis also quickly discovered that scarcely any of the new technology developed in recent years was being used to produce windmills.

Economic and political winds were changing fast when this happened. ARCO stopped running television commercials belitting Danish windmills. The Shell Oil Company was about to evaluate the wind's usefulness as a source of electrical current on production platforms in the Gulf of Mexico. Professorial proponents of windpower were stirring up more interest in it in Oklahoma, Minnesota, Oregon, and other states, and Culebra was not the only island with more wind than fossil fuel.

The Block Island Power Company was planning to use wind-driven generators on the isle it served, and the Boston Edison Company encouraged a similar venture in its territory. The Federal Power Commission discovered "a genuine interest on the part of all electric utility companies in economical but non-polluting sources of power." And Joe Savino's review of both Palmer Putnam's experience and Percy Thomas's conclusions convinced him that, with a few hundred thousand more dollars, the S. Morgan Smith Company might have developed a power plant that would be helping utility companies meet peak loads economically today.

Although the cost of research as well as fuel had risen since then, the National Science Foundation supported the work begun at the Lewis Research Center and elsewhere. For a time it seemed as though government agencies were jumping on their horses and racing off in all directions. The Department of Agriculture and the National Park Service had long been interested in windmills, and the Federal Energy Administration supported studies of "energy parks" in which clusters of industries might be served by more than one kind of power plant. When ERDA was organized, it absorbed the Atomic Energy Commission, took over some of the National Science Foundation's work, and became the principal source of government funds for energy research.

### THE TESTING POWER

ERDA's first wind energy conversion system is the biggest one in the United States in 1976. It is on a smooth field in NASA's Plum Brook facility near Sandusky, less than a mile from the Ohio Turnpike. This site was chosen for visitors' convenience rather than because additional kilowatts were needed there any more than at other places. Even more engineers are expected to study this system than studied the giant on Grandpa's Knob. Although its completion received little publicity, 532 persons drove out to see it the first month that the site was opened to the public four hours a day five days a week.

The tower at Plum Brook is only nine-tenths as high as the one torn down in Vermont, and the generating system is less than one-tenth as powerful. It produces a maximum of 100 kilowatts, which would be enough for only about 25 homes today. This rig is worth watching, however, because it is being used to test new concepts, components and the construction methods that have been proposed to reduce the cost of building much more powerful wind turbines.

Most visitors have probably already seen similar open truss towers; utility companies have built them for transmission lines. But a cabin that resembles an airplane's fuselage is perched on this one, and ten flights of stairs in the tower enable engineers and technicians to go up and inspect or alter the generating machinery in that tubular housing.

A propeller with two blades that look like the wings of an airplane is attached to one end of the cabin pivoted on the tower. Each blade is 4 feet wide at the hub, only 18 inches at the tip, and twisted about 26 degrees. Both blades are oyster white with tangerine bands on them, and each one weighs about 2,000 pounds. These are aluminum blades with no welds in them for the wind to crack. When they have been adequately tested, airfoils made out of composite materials that would be cheaper to manufacture and more durable may be substituted for them.

The average wind velocity at Plum Brook is about 10 m.p.h., south-southwest in summer and north-northwest in winter; the wind exceeds 18 m.p.h. for long enough periods every month for the tests under way. Sensors and servomotors keep the cabin aligned with the wind and the propeller on the downwind side of the tower. The generator begins to produce power when the wind exceeds 8 m.p.h., and the propeller revolves 40 times a minute when the wind is blowing between 18 and 40 m.p.h. The blades are automatically feathered whenever the wind becomes ferocious, and it frequently has. The pitch-changing mechanism for the blades is unique, but similar to governors developed for airplane propellers and helicopter rotors.

Aeronautical technology, in fact, was used throughout the construc-

*NASA/ERDA experimental wind generator ready for tests at Plum Brook.*

tion of this experimental plant. The Lockheed company built it for NASA, and the generating system delivers 480-volt, 3-phase, 60-cycle current that could be used in today's utility systems. It is a small plant compared to the one the S. Morgan Smith Company built in the 1940s, but Louis Divone, a 1956 M.I.T. graduate employed by ERDA, has explained that it was "about the largest size we felt we could start out with."

Allis-Chalmers bought the S. Morgan Smith Company in 1949, but its former president, Beauchamp E. Smith, and one of its engineers, Carl Wilcox, are still active. They were special guests at the formal dedication of the Plum Brook plant. "When we came through the gates today," Smith told a reporter, "I got a feeling of great satisfaction. I've always felt that something good would come out of our tests even though we were ridiculed at the time, when energy sources seemed to be more abundant than our country would ever need. We were just ahead of our time." And Wilcox added, "After thirty-five years our ideas have been proved worthwhile."

From Plum Brook's 100-kilowatt plant, ERDA expected to "scale up" by 1981 to power plants rated at 3,000 kilowatts—twice as high as the Smith-Putnam generator. Those units will still not be in the same ball park with fossil-fuel and nuclear plants that the utilities have today. A thousand big wind turbines may be needed to equal the output of a single nuclear plant. ERDA has continued, therefore, to invest more millions in nuclear research and development than in efforts to catch the wind. But ERDA is trying, the way Casey Stengel might have, to "cover all bases" in the crucial energy game.

Small power plants will continue to be needed in many communities no matter what is done elsewhere. Fuel may still be expensive in many parts of Alaska, for instance, when the big new oil pipeline is completed there. To begin studies of small wind turbines, NASA bought a French-built Aerowatt, rated at only 4.1 kilowatts, and put it up alongside the new 100-kilowatt machine.

Dr. John M. Teem, a former assistant administrator of ERDA, once predicted that 7 percent of our country's energy will come from solar sources, including the wind, by the year 2,000. The wind will not be neglected, for it often blows when the sun does not shine. The huge hydro-electric plants that we have now produce only about 3 percent of the electric current we consume today, and some could produce more kilowatts if windpower were used to replenish the water in the reservoirs behind the dams.

### STRAWS IN THE WIND

When I asked my broker recently about stock in a new company organized to produce small aerogenerators, she was startled. The price had tripled in less than a year, but she warned me that this could be a straw in the wind of no significance. What happens at Congressman Henry R. Reuss' summer home at North Lake, Wisconsin, might be much more indicative of the future course of events.

*Congressman Henry Reuss posing with the 2-kilowatt wind plant on his farm.*

The Wisconsin Windworks researchers put a small aerogenerator on an octagonal tower at North Lake to test a little black box envisioned long ago by Palmer Putnam on Cape Cod. It was called a "Gemini Power Conversion Unit," and converted the direct current that the wind generated into alternating current at a voltage and frequency that could be fed into the utility company's lines. The congressman could make his electric meter run backwards, and thereby reduce his bill, whenever the wind generator produced more current than he was using.

Congressman Reuss, however, permitted the power company to put a ratchet on the meter to prevent this, because he realized that widespread use of such devices would necessitate revision of the Wisconsin Public Service Commission's rules and rate structures. If successful, such converters could reduce many families' bills. Storage batteries would no longer be needed with home aerogenerators, and those converters might also help the utility companies meet peak loads economically.

"As you can see," the congressman wrote Robert C. Seamans, Jr., ERDA's administrator, "our windmill contribution is entirely a private-sector contribution. I would be most happy if ERDA could take the experiment from here out, and would be pleased to make available my physical plant to you." A hint from a powerful committee chairman in the House of Representatives could lead to a sequence of events similar to that which a question from the Governor of Puerto Rico started a few years ago. ERDA must consider the nation's security as well as the cost of its energy—and a hundred thousand small wind turbines widely scattered would be much harder for an enemy to knock out than any kind of big central power plant.

The big wind plants proposed, of course, will be much more spectacular sights than little black boxes converting direct to alternating current. Some may really amaze nostalgic windmill buffs. One concept that the McDonnell-Douglas Corporation was analyzing on a computer in 1976 was for a huge "giromill." That name was derived from "cyclogiro" and "windmill"—but, if built, a giromill may look more nearly like a riverboat's paddlewheel than an Autogiro. The "paddles" envisioned would be enormous airfoils. The wind might whirl four of them, 130 feet tall, around an upright mast to generate current for the utility companies to distribute.

Alberto Kling, a distinguished and dynamic Danish authority on wind-driven devices, was invited to the 1975 wind energy workshop in Washington. He brought pictures from Denmark showing how vertical-axis wind turbines could be mounted attractively, either separately or stacked in a tall tower. Kling also displayed a model of a "precessor system for wind energy conversion" that is being studied in his country. It both startled and puzzled many participants in the workshop.

After Kling addressed a formal session of the workshop, NASA's Joe Savino pulled a folded piece of paper out of his pocket. When he unfolded it, he had a row of silhouettes of a tiny windmill. Savino bent it into a crown to place on Kling's head. Then he inquired seriously about plastering new instruments all over a big Danish machine at Gedser, which had been standing idle since 1968, to obtain more data from it.

While I was staring at Kling's model of a "precessor system," a young engineer bent over to examine it more closely. When he looked up he grinned and asked me, "Wouldn't it be funny if we got into a technological race with the Danes to catch the wind—the way we got into one with the Russians to get to the moon first?" Developing big economical and highly efficient wind-catchers may prove to be as great a challenge as sending a man to the moon and back. Few Americans ever expected to go to the moon; many more want the cost of energy reduced, and many ways of doing it may have to be tested.

### QUESTIONS AND ANSWERS

What new version of the sturdy old windmills will be most efficient, attractive, and inexpensive to construct?

How long will it take to get all of the bugs out of big wind energy conversion systems, when inexperienced men restrained by lawyers and public relations experts take charge of those new power plants?

Will small aerogenerators adequate for a modern home be manufactured sooner? And will they be simple enough for a home handyman to assemble and erect?

These are good questions that can be clarified in the next quarter of our century. Since 1972 the United States Patent Office has given top priority to energy-related devices. Even if the public's fears of nuclear power are fully dissipated, some people will continue to build windmills. In many places the wind is the most accessible of the earth's many randomly distributed resources. And wind engines have always been as alluring as those newly discovered nuclear particles which have a puzzling property that the physicists call "charm."

When Noah sat down to dine in his ark long ago, he is thought to have said that he did not care where the water went, as long as it did not get into the wine. Until recently, few of us cared where the smoke, soot, and other debris from our power plants went, as long as they did not interfere with our pleasures. But the earth's sweet aeolian breath has begun to sour, and many people care now. Wind-driven engines have never increased the need for deodorants, and the best windmanship may be developed where the earth has been most polluted.

The Wind Energy Society of America, organized in California, hopes to help today's researchers exchange information more swiftly. So does an American Wind Energy Association with headquarters in Michigan. Journalists, too, are trying to inform the public about the new WECSes, the way they have enlightened some of us about sexes. This book has not told you all that you should know about windmills before you build or buy one, but it probably has told you more than you knew before about the history of these fascinating machines.

Suppose that you are a gentleman and have met a pretty lady whom you would like to know better. You have discussed the weather, your health, and your jobs, and come to an impasse. What can you say to her next? You can always ask what she knows about windmills. Everybody remembers at least one, and if the lady is better informed on the subject than you are, so much the better.

If she mentions Don Quixote, she is interested in the arts. If she refers to Willa Cather or Mari Sandoz, you will know she is interested in the Great Plains. If she has seen windmills in Holland, she has been to Europe. If she mentions a stamp with a picture of a windmill on it, she collects stamps. And if she thinks you have windy people in mind, she may be a psychologist. Whatever she replies, you can be conversationally airborne with her again about something or other.

Have a nice tomorrow!

# APPENDIX I

# *For More Information*

### General Works on Sources of Energy

*Energy for Survival,* by Wilson Clark. New York: Anchor Press/ Doubleday, 1974. An up-to-date survey now used as a text in some schools. It includes a lengthy discussion of windpower. Both hardbound and paperback editions are available.

*The Energy Crisis,* by Lawrence Rocks and Richard P. Runyon. New York: Crown Publishers, Inc., 1972. An older textbook with less emphasis on windpower.

*The Energy Balloon,* by Stewart Udall. New York: McGraw-Hill, 1974. An especially interesting account of the dimensions of the problem.

### Help for Home Handymen

Anyone planning to build all or any part of a windmill for himself should consult a neighbor who has one if possible. Manufacturers' salesmen also can be helpful; few of them will remind you of used car salesmen.

*Electric Power from the Wind,* published and sold by Henry Clews. East Holden, Maine 04429, 1973.

*Wind and Windspinners,* by Michael A. Hackleman. Available from Earthmind, 20510 Josel Drive, Saugus, California, 91350 1974.

*Alternative Sources of Energy.* A bi-monthly newsletter published at Rt. 1, Box 36 B, Minong, Wisconsin, 54850.

*Energy Book #1,* John Prenis ed. Philadelphia: Running Press, 38 South Nineteenth St., 19103, 1975.

*Wind Power Digest.* Available from Mike Evans, Rt. 2 Box 489, Bristol, Indiana 46507.

## Periodicals on Current Developments

*Popular Science Monthly*, 380 Madison Ave., New York 10001. A widely available source of reliable news of innovations in the use of the wind. Many public and other libraries file it and every issue is indexed.

*Science*, published weekly by the American Association of Science, 1615 Massachusetts Ave., NW, Washington, D.C., 20005. Covers both research and government policies affecting all sources of energy.

(Many professional journals, popular periodicals, and newsletters issued by trade and other associations emphasizing energy, the environment, and ecology, also keep readers abreast of developments affecting wind-powered devices.)

## For Up-To-Date Reference

*Wind Energy Utilization.* A bibliography prepared for the NASA Lewis Research Center by the Technology Application Center of the University of New Mexico at Albuquerque (cumulative volume 1944/ 1974, TAC W 76-700) is one of the most complete and up-to-date available.

*Wind Energy Bibliography.* Published in 1973 by Windworks, Box 329, Rt. 3, Mukwonago, Wisconsin, this is a shorter listing of both older and recent additions to the literature.

## References for Chapter Subjects

### CHAPTER I

REYNOLDS, John: *Windmills and Watermills.* London: Hugh Evelyn, 1970.
DELITTLE, R. J.: *The Windmill Yesterday and Today.* London: John Baker, Ltd., 1972.
"Power From the Wind." *Engineering and Science*, Published by the California Institute of Technology, May 1974.

### CHAPTER II

IRVING, Washington: *Mahomet and His Successors.* New York: G. P. Putnam's Sons, 1850.
CARLYLE, Thomas: *Past and Present.* Boston: Houghton Mifflin, 1965.
USHER, A. P.: *A History of Mechanical Invention.* Cambridge, Mass.:

Harvard Press, 1929; Beacon Press paperback, 1959.

WHITE, Lynn, Jr.: *Medieval Technology and Social Change*. Oxford University Press, 1962.

MUMFORD, Lewis: *Art and Technics*. New York: Columbia University Press, 1952.

### CHAPTER III

BARBOUR, Erwin Hinckley: *The Homemade Windmills of Nebraska*. Bulletin 59, Agricultural Experiment Station of Nebraska, Vol. XI, Art. V.

BATHE, Greville: *Horizontal Windmills, Draft Mills, and Similar Air-Flow Engines*. Philadelphia: 1948.

LEY, Willy: *Engineers' Dreams*. New York: Viking Press, 1954.

WULFF, Hans: *Traditional Crafts of Persia*. Cambridge, Mass.: M.I.T. Press, 1971.

### CHAPTER IV

BENNETT, Richard, and ELTON, John: *History of Corn Milling, Vol. 2, Watermills and Windmills*. Wakefield, Yorkshire, England: EP Publishing, Ltd., 1972.

FREESE, Stanley: *Windmills and Millwrighting*. Cranbury, N.J.: Great Albion, 1957.

VINCE, John: *Discovering Windmills*. Aylesbury, England: Shire Publications, Ltd., 1973.

PRESTON, Dickson J.: "One Man's Answer to the Energy Crisis." *Old Mill News* Vol. 1, No. 4, July 1973.

*The Miller in Eighteenth Century Virginia*. Williamsburg Craft Series Publications, 1957.

### CHAPTER V

STOCKHUYZEN, Frederic: *The Dutch Windmill*. New York: Universe Books, 1963.

SPIER, Peter: *Of Dikes and Windmills*. New York: Doubleday, 1909.

VAN VEEN, Johan: *Dredge Drain Reclaim!* The Hague: Martinus Nijhoff, 1962.

HAMILTON, Roger: "In Praise of Windmills, Victims of a World They Helped Create." *New York Times*, April 29, 1973.

GOULD, Donald: "Draining the Zuider Zee uncovers a boneyard of ancient ships." *The Smithsonian*, March 1974.


CHAPTER VI

WAILES, Rex: *The English Windmill*. London: Routledge & Kegan Paul Ltd., 1971.

SKILTON, C. P.: *British Windmills and Watermills*. London: Collins, 1947.

FARRIES, K. G. and MASON, R. T.: *The Windmills of Surrey and Inner London*. London: Chas. Skilton, Ltd., 1966.

SHORT, Michael: *Windmills in Lambeth*. Published by Director of Libraries and Amenities, Tate Central Library, London, 1971.

SMEATON, John: *An Experimental Enquiry concerning the Natural Powers of Water and Wind to turn Mills and other Machines, depending on a Circular Motion*. London: I and J. Taylor, 1794.

MAYR, Otto: *The Origins of Feedback Control*. Cambridge, Mass.: M.I.T. Press, 1970.


CHAPTER VII

ZEMURRAY, Martha and Murray: *Early American Mills*. New York: Clarkson N. Potter, 1973.

BISHOP, J. Leander: *A History of American Manufacturers*. Vol. 1. Edward Young & Co., 1861.

WAILES, Rex: *Windmills of Eastern Long Island*. Port Washington, N.Y.: Ira J. Friedman, Inc., 1962.

JORDAN, Terry G.: "Evolution of the American Windmill." *Pioneer America* Vol. 5, No. 2, July 1973.

AMORAL, Theodore F.: "The Old East Mill at Heritage Plantation." *Old Mill News* January, 1974.

DUNBAR, Gary S.: *Historical Geography of the North Carolina Outer Banks*. Baton Rouge: Louisiana University Press, 1958.

WEISS, Henry B. and Grace W.: *Early Windmills of New Jersey*. N.J. Agricultural Society, 1969.

WRENN, Tony P.: Beaufort, N.C., study for division of archives and history. Dept. of Cultural Resources, State of North Carolina, Raleigh, December 1970.

ARMBRUST, Henry W., et al: *The Old Jamestown Mill*. Jamestown, R.I., Historical Society, 1964.

CHAPTER VIII

WEBB, Walter Prescott: *The Great Plains*. Lexington, Mass.: Ginn & Co., 1961.

WOLFF, Alfred R.: *The Windmill as a Prime Mover*. New York: J. Wiley & Sons, 1885.

*Pioneers in Industry, 1830–1945*. A history of Fairbanks, Morse & Co., published by the company.


CHAPTER IX

EIDE, Clyde A.: "Free as the Wind." *Nebraska History* Spring 1970.

DICK, Everett: *The Sod-House Frontier, 1854–1890*. New York: D. Appleton-Century, 1939.

ANDERSON, Bob: "Early Nebraska Partners, Water and Wind." *Rural Electric Nebraskan* April 1972.

GRABB, John R.: "The Windmill and Pump: They Made the Great Plains Blossom." *The Chronicle of Early American Industries Association*, April 1974.

McCAULEY, Ruth: "Windmills and Windmill Weights." *Spinning Wheel* March 1967.


CHAPTER X

MURPHY, Edward Charles: *The Windmill: Its Efficiency and Economic Use*. U.S. Geological Survey Paper No. 41–42. Washington, D.C., Government Printing Office, 1901.

FETTERS, Jim: "Windmills: Phenomena in the Atomic Age." *Water Well Journal* February 1972.

Articles in *Omaha World-Herald*, Feb. 4, 1962; *Lincoln* (Neb.) *Star*, May 29, 1973; *Kansas City Star*, July 23, 1964, and *New York Times*, Dec. 21, 1972.


CHAPTER XI

JACOBS, Marcellus: Interview in *Mother Earth News* November 1972.

TRUCK, Ed: "Free Power From the Wind." *Mother Earth News* September 1972.

McCOLLEY, H. F. and BUCK, Foster: *Homemade Six-Volt Wind-Electric*

*Plants*. Special Circular from the North Dakota Agricultural College Extension Service. Fargo: January 1939.

UNESCO *Conference in Rome on New Sources of Energy*, 1961. Economic and Social Council Report. February 1962.

GOLDING, E. W.: *The Generation of Electricity by Windpower*. London: E. & F. N. Spon, Ltd., 1955.


## CHAPTER XII

PUTNAM, Palmer Coslett: *Power From the Wind*. New York: Van Nostrand, 1948.

KARMAN, Theodore von (with Lee Edson): *The Wind and Beyond*. Boston: Little Brown & Co., 1967.

HOWARD, Lawrence M.: "Power from the Winds." *Vermont Life*, Vol. 10, Winter 1955–6.

McCAULL, Julian: "Windmills." *Environment*, Vol. 15, No. 1 January/February 1973.

THOMAS, Percy H.: *Aerodynamics of the Wind Turbine, Electric Power from the Wind, Fitting Wind Power to the Utility Network, and The Wind Power Aerogenerator*. Prepared for Office of Chief Engineer, Federal Power Commission, 1946–1959.


## CHAPTER XIII

JESPERSEN, Anders, ed.: *Transactions of the Second International Symposium on Molinology, May 1969*; and VAN HOOGSTRATEN, M. (compiler): *Transactions of the Third Symposium on Molinology*, in The Netherlands, 1973. (Stephen M. Kindig, The Gristmill at Lobachsville, Oley R.D. 2, Pennsylvania 19547, represents TIMS, the sponsor of these symposia, in the United States.)

GRIEG, Michael: "Energy Plea for Park Windmills." *San Francisco Chronicle*, Dec. 29, 1973; and an unsigned article about the Golden Gate windmills in *Old Mill News*, Vol. III, No. 1, January 1975.

WALTON, James: *Water-mills windmills and horse-mills of South Africa*. C. Struik Booksellers, P.O. Box 1144, Cape Town 8001, 1974.


## CHAPTER XIV

SWEENEY, T. E.: *The Princeton Windmill Program*. Princeton University

Department of Aerospace and Mechanical Sciences, AMS Report No. 1093, March 1973.

MEYER, Hans: "Wind Generators, Here's an Advanced Design You Can Build." *Popular Science Monthly* November 1972.

HESS, Karl: "Washington Utopia." *Washington Post* Nov. 3, 1974.

BARNHART, Earle: "Wind Power." *The Journal of The New Alchemists.* New Alchemy Institute, Woods Hole, Massachusetts 02543, 1973.


CHAPTER XV

HERONEMUS, William E.: "The U.S. Energy Crisis, Some Proposed General Solutions." *Congressional Record*, Vol. 118, No. 17—Part II, Feb. 9, 1972.

HERONEMUS, William E.: "Using Two Renewables." *Oceanus*, Vol. XVII, Summer 1974.

WOLCOTT, John: "William Heronemus, Unlikely Revolutionary." *Country Journal*, November 1974.

PETERSON, John: "Energy Blowing in the Wind." *National Observer*, May 3, 1975.

CROWE, Bernard J.: *Fuel Cells, a Survey.* NASA Publication SP-5115. Government Printing Office, 1973.

SAVINO, Joseph, ed.: *NSF/NASA Wind Energy Workshop Proceedings, June 11–13, 1973.* National Science Foundation Publication RA/W 73-006, December 1973.

BAMBERGER, C., and BRAUNSTEIN, J.: "Hydrogen: A Versatile Element." *American Scientist* July–August 1975.


CHAPTER XVI

THALLER, Lawrence H.: *Electrically Rechargeable Redox Flow Cells.* A paper prepared for an Intersociety Energy Conversion Engineering Conference in San Francisco, Aug. 26–30, 1974. NASA Technical Memorandum TMX-71540, Lewis Research Center.

BLACKWELL, B. F.: *The Vertical-Axis Wind Turbine "How It Works."* Sandia Laboratories Energy Report, April 1974. Publication SLA 74-0160.

HACKLEMAN, Michael: "More About the S-Rotor." *Mother Earth News* May 1974.

VILLECO, Marguerite: "Wind Power." *Architecture Plus*, May/June 1974.
STEPLER, Richard: "Eggbeater Windmill." *Popular Science Monthly* May 1975.

## CHAPTER XVII

DEKORNE, James B.: "The Answer is Blowin' in the Wind." *Mother Earth News* November 1973.

CARTER, Joe: "Wind Power for the People." *Organic Gardening and Farming* August 1975.

ROCKER, Winnie Red: "Build a Wind Generator!" *Alternative Sources of Energy* January 1973.

GREGOR, Sandra L.: "Wind Power is a Viable Alternative to Nuclear Power." *Maine Times* Vol. 5, No. 30, May 4, 1973.

MERRIAM, Marshal F.: *Is There a Place for the Windmill in the Less Developed Countries?* Working Paper Series No. 20, March 1972, Technology and Development Institute, The East-West Center, Honolulu, Hawaii 96822.

## CHAPTER XVIII

SAVINO, Joseph M.: *A Brief Summary of the Attempts to Develop Large Wind-Electric Generating Systems in the U.S.* A paper presented at a conference sponsored by the Swedish Board for Technical Development, Stockholm, August 29–30, 1974. NASA Technical Memorandum TMX-71605, Lewis Research Center.

THOMAS, R., PUTHOFF, P., and SAVINO, J.: *Plans and Status of the NASA-Lewis Research Center Wind Energy Project.* A paper prepared for a joint meeting of the Institute of Electrical Engineers and American Society of Mechanical Engineers, Portland, Oregon, Sept. 28–Oct. 1, 1975. NASA Technical Memorandum TMX-71701, Lewis Research Center.

LINDSLEY, F. E.: "Wind Power." *Popular Science Monthly* July 1974.
*Information from ERDA* (press releases). Vol. 14, 1975.

# APPENDIX II

# Wind Engine
# Manufacturers and Associations

### American Manufacturers

Fan-type farm windmills for running pumps are manufactured by:

*The Heller-Aller Co.*, Napoleon, Ohio 43545.
*Dempster Industries, Inc.*, P.O. Box 848,
Beatrice, Neb. 68310.

Until recently, the only manufacturer of nationally advertised aerogenerators in the United States was:

*The Winco Division, Dyna Technology, Inc.*,
East Seventh at Division Street, P.O. Box 3263,
Sioux City, Iowa 51102.

Some new firms offering or hoping soon to offer aerogenerators for sale are:

*American Energy Alternatives, Inc.*, P.O. Box 905,
Boulder, Colorado 80302.
*American Wind Turbine, Inc.*, 1016 Airport Road,
Stillwater, Oklahoma 74074.
*Wind Power Systems, Inc.*, P.O. Box 17323,
San Diego, California 92117.
*Helion*, P.O. Box 4301, Sylmar, California 91342.
*North Wind Power Company.*, Warren, Vermont 05674.

Many more firms are affiliated with the American Wind Energy Association, 21243 Grand River, Detroit, Michigan 48219.

### Foreign Manufacturers

Dealers in several states have imported windmills and aerogenerators made abroad. Some of the best known foreign makers are:

217 ࿔

*Dunlite Electrical Co.,*
Division of Pye Industries,
21 Fromme St.,
Adelaide 5000 Australia.

*Aerowatt S. A.,*
37 Rue Chanzy,
75 Paris lle, France.

*Elektro G. m. b. H.,*
Winterthur, St. Gallerstrasse 27 Switzerland.

*Lubing Maschinenfabrik,*
Ludwig Bening,
2847 Barnstorf (Bez Bremen)
Postfach 171, West Germany

## Windmill Plans

Plans for building windmills of various kinds may be obtained from:

*Brace Research Institute,* MacDonald College, McGill University, Ste. Anne de Bellevue 800, Quebec, Canada.

*VITA,* Volunteers for Technical Assistance, 3706 Rhode Island Ave., Mount Rainier, Maryland 20822.

*Sencenbaugh Wind Electric,* P.O. Box 11174, Palo Alto, California 94306; or 678 Chimalis Ave., Palo Alto, 94301.

*Windworks,* Box 329, Rt. 3, Mukwonago, Wisconsin 53149.

## Windmill Associations

Local newspapers and libraries are likely to be the best sources of information about old windmills in particular localities. More books and booklets have been published about the early English smock mills than any other type used in this country. New England libraries have shelves full of books that describe windmills north of Long Island Sound. The East Hampton Free Library's Gardiner Memorial Room has detailed records of many Long Island windmills.

Addresses of some lively organizations interested in preserving windmills follow:

*International Molinological Society (TIMS),*
J. Kenneth Major, secretary, 2 Eldon Road,
Reading RGI 4 DH, England.

*Manufacturers and Associations*

*Society for the Preservation of Old Mills,*
Everett S. Powers, business manager,
232 Roslyn Ave., Glenside, Pennsylvania 19038

*Society for Industrial Archeology,*
Robert M. Vogel, editor, Room 5020
National Museum of History and Technology,
Smithsonian Institution, Washington, D.C. 20560

*Windmill Preservation Association* (WPA),
E. E. McGregor, secretary, P.O. Box P,
Gibbon, Nebraska 68840.

*McLaren Windmill Fund,*
Eureka Federal Savings and Loan Co.,
4610 Mission St. San Francisco, California 94110.

# Index

# Index

# Index